THE POLITICS OF POWER
ONTARIO HYDRO AND ITS GOVERNMENT, 1906–1995

Ontario Hydro is a paradox. Omnipresent and omnipotent in the Ontario political and economic landscape, its nature and identity have been shrouded in ambiguity for ninety years. *The Politics of Power* provides a fascinating account of Hydro's origins and history up to the 1995 provincial election.

Freeman contends that the common perception of Hydro as the archetypal crown corporation is mistaken, despite its reputation as one of the first and most important examples of large-scale public enterprise in Canada. From the legislation that launched the Hydro-Electric Power Commission of Ontario (HEPC) in 1906 to its formal re-creation as Ontario Hydro in 1973, the utility was simultaneously considered in different quarters to be both a government enterprise and the trustee of a municipal cooperative.

This ambivalence continues to be a central theme in Hydro's history. As Freeman shows, the ownership confusion was only attenuated rather than terminated with the creation of Ontario Hydro, and this has implications for its restructuring and privatization today. While municipal ownership is largely a myth, it has survived so long not only because municipal leaders gave it articulation; it conveniently supported the political objectives of Hydro to bolster corporate autonomy and the government to silence criticism of direct involvement in the economy.

Through meticulous examination of statutory changes and government appointments, and through candid interviews with key government, municipal, and Hydro officials, Freeman gives us a much clearer understanding of this important corporation and its government.

NEIL B. FREEMAN is a public policy consultant and an adjunct professor in the Department of Political Science at the University of Toronto.

NEIL B. FREEMAN

The Politics of Power: Ontario Hydro and Its Government, 1906–1995

UNIVERSITY OF TORONTO PRESS
Toronto Buffalo London

Toronto Buffalo London
Printed in Canada

ISBN 0-8020-0798-8 (cloth)
ISBN 0-8020-7160-0 (paper)

Printed on acid-free paper

Canadian Cataloguing in Publication Data

Freeman, Neil B. (Neil Bryce), 1958–
The politics of power

Includes bibliographical references and index.
ISBN 0-8020-0798-8 (bound)
ISBN 0-8020-7160-0 (pbk.)

1. Ontario Hydro – History. 2. Electric utilities – Government ownership – Ontario – History. I. Title.

HD9685.C3405 1996 333.79'32'060713 C95-932895-5

University of Toronto Press acknowledges the financial assistance to its publishing program of the Canada Council and the Ontario Arts Council.

For my mother and father

Contents

Preface

Ontario Hydro is a paradox. In the Ontario political landscape, it is always omnipresent and sometimes omnipotent, yet little is known about its character and relationship with the Ontario government. With concerns as wide-ranging as rate determination and sustainable development, the paradox is as baffling for consumers as it is for political observers. This absence of knowledge has been made all the more notable with the election of the Progressive Conservatives in 1995, who won with talk of privatizing Ontario Hydro after eighty-nine years of public existence as part of their platform. By unearthing the background to such contemporary issues, *The Politics of Power* seeks to narrow this gap in knowledge.

I came to explore Ontario Hydro's public ownership and relationship with the government after having intended first to study its nuclear power policy from 1973 to 1982. Although this period had all the makings of a fascinating story – beginning with a massive expansion of nuclear capacity to meet the energy crisis and ending with the slowdown of construction in the midst of a prolonged recession – the unfurling of the tale was stymied. The lack of a general understanding of Hydro's relationship with the Ontario government stood in the way. My investigation of this problem led me to conclude that this relationship was a book in itself, one not explained by the extensive literature on crown corporations in Canada. As the new project progressed, I began to realize that it contained two distinct parts. The book contained here details the first part – from the inception of the Hydro-Electric Power Commission of Ontario (HEPC) in 1906 to its formal re-creation as Ontario Hydro in 1973 – and records the second part – the politics of the Ontario Hydro since 1973 – through an epilogue.

Although Ontario Hydro currently appears as the archetype of a government enterprise, the HEPC's public ownership and relationship with the government were strikingly different from this organizational form. It exhibited the ambivalence of both a government enterprise and the trustee of a municipal cooperative. While municipal cooperative ownership was only an elaborate myth disguising government ownership, the contention was nevertheless made powerful through statutory enactments and government appointments. Indeed, the competing ownership views stood continuously unresolved because of these mechanisms, and the confusion was not completely ended with the creation of Ontario Hydro in 1973. This fact has had a profound impact on the politics of Hydro's relationship with the Ontario government to this day.

The ownership ambiguity of the HEPC is a subject almost forgotten from Ontario memory. Indeed, the history of the commission has been absorbed seamlessly into that of Ontario Hydro, with the reforms of 1973 remembered as little more than a name change. While the physical assets of the utility did not change with the transition from a commission to a corporate organizational structure, a profound change nonetheless occurred in the understanding of the nature of the organization. Until the creation of Ontario Hydro it was quite common for the provincial and municipal hydro commissions to be jointly thought of as 'the hydro,' with lines of authority to the provincial and municipal governments simply illustrative of a joint governmental contribution to a single pattern of public delivery of electric power. In the postwar period the ownership of the HEPC increasingly became the subject of provincial-municipal conflict, with the re-creation of the commission as Ontario Hydro only the most concerted attempt for a definitive resolution. Although the HEPC was informally known as Ontario Hydro, and both are commonly referred to as simply 'Hydro,' this book keeps the names distinct so that the peculiar ownership claims over the HEPC and the significance of the creation of Ontario Hydro can be seen in their full light.

The researching and writing of this book relied extensively on primary documents. I was fortunate in that most of what I wanted to examine was on the public record. To this end, a number of people provided valuable reference assistance. I would particularly like to thank the following: Silvia Ernasaks, Glenys Biggar, Joanne Collingwood, and Mary Wagner of the Ontario Hydro Public Reference Centre; Trish Wilcox of the Ontario Hydro Archives; Leon Warmski, Karen Bergsteinsson, and Jim Suderman of the Archives of Ontario; Sheree Bond and Maurice Tucci of the Municipal Electric Association; and the desk staff of the invaluable and voluminous government documents section of the Robarts Library, University of Toronto.

My primary research also included interviews and correspondence I conducted with senior Ontario Hydro, Ontario government, and Ontario Municipal Electrical Association officials. I thank them all for their assistance; my research would not have been complete without their contribution. With names too numerous to mention here, a complete list is included in the bibliography. I would like to note at this point, however, that I was pleased to be able to meet George Gathercole, very much the architect of Ontario's postwar economic project, before he died in late 1994.

The rich work of the historians who have generated the not inconsiderable secondary literature for researchers on Ontario subjects made the primary research somewhat less onerous. I am especially indebted to H.V. Nelles and the authors in the Ontario Historical Studies Series for having trodden some of this ground before me. Indeed, reading Nelles's *The Politics of Development* as an undergraduate stimulated my original interest in Ontario Hydro.

Gratitude must also be expressed to a number of academics and insiders for having read the various chapters of the book in preliminary form. I was fortunate to receive comment on the first substantive chapter, number two, from H.V. Nelles, Department of History, York University; Jean Laux, Department of Political Science, University of Ottawa; Allan Tupper, Department of Political Science, University of Alberta; and Kent Weaver, The Brookings Institution, Washington, DC. The core of the manuscript was read by Douglas Gordon, formerly president of Ontario Hydro; Gordon McHenry, formerly general manager of personnel of Ontario Hydro; Tom May, general manager, Whitby Hydro; and Keith Reynolds, formerly deputy minister to the premier and secretary to the cabinet, Government of Ontario. The whole book received the scrutiny of J.E. Hodgetts, Skelton-Clark Fellow in Political Studies, Queen's University, and of another reviewer who remains anonymous. I thank all of them for their assistance, which I can unreservedly say improved the final product.

The publication of this book was financially assisted by a subvention from the University of Saskatchewan's publication fund. I received the benefit of this generosity while spending a most enjoyable year as a member of that institution's Department of Political Studies. The compilation of the index was supported by Wilfrid Laurier University's book preparation fund. An early version of chapter two was published in *Ontario History* and I wish to express my thanks for permission to include it here.

Like many an academic author's first book, this one began as a doctoral thesis. I would be remiss in not paying tribute to the terrific cast of supervisors and examiners I had when writing and defending the thesis. Beginning with the latter, I had the benefit of the oversight of Ken Rea, Department of Economics, University of Toronto, and Allan McDougall, Department of Political Science, University of Western Ontario. As for my supervisory committee, all of the Department of Political Science, University of Toronto, I owe a great debt to Marsha Chandler (also dean of Arts and Science), Graham White, and Peter Silcox for their assistance. Special mention must be made of the contribution of Stefan Dupré, my supervisor. Without the guidance of his great intellect and rigorous standards of scholarship, the thesis would be much less of a contribution to knowledge.

For recognizing that the thesis would make a good book, I am indebted to Virgil Duff, executive editor, University of Toronto Press, who went beyond the call in expediting its publication. Near the final completion of the book I was fortunate to have Catherine Sims give the manuscript her detailed historical and editorial scrutiny. Coming to the aid of the general reader, my dear friend Patricia Phillips then applied her editorial talents to excising my sometimes awkward academic prose. Doug Hamilton ably compiled the index. The copy editing was performed by Diane Mew, and I am happy to have it in the esteemed company of the many works to which she has given polish. While I am thankful for the assistance of all the above, any errors and omissions are mine alone.

I have dedicated this book to my mother, an Ontarian to the core, and my father, an Australian by birth and temperament, for a reason. Without the chemistry this meeting of minds provided, my interest in political matters would not have been as acute.

NEIL FREEMAN
Toronto
July 1995

THE POLITICS OF POWER

1

Introduction

Ontario Hydro's character and relationship with the Ontario government have had a distinctiveness all their own. Although Hydro was among the first and remains one of the foremost examples of large-scale public enterprises in Canada, the common perception that it has always been the archetype of a crown corporation is mistaken. This is especially the case for its incarnation as the Hydro-Electric Power Commission of Ontario, where the most striking feature was something quite different. From the legislative episode which launched the HEPC in 1906 to its formal re-creation as Ontario Hydro in 1973, it was simultaneously considered in different quarters to be both a government enterprise and the trustee of a municipal cooperative. Indeed, this 'institutionalized ambivalence,'[1] to use Carolyn Tuohy's axiom of Canadian politics and public policy, was a central feature of its history. Despite numerous attempts to seek a resolution, the confusion over the HEPC's public ownership was only attenuated rather than terminated in the establishment of its successor. Ontario Hydro's continued restructuring and possible privatization remain complicated as a result.

Although municipal cooperative ownership was, in fact, a myth disguising government ownership, the contention was nevertheless powerful. As an outgrowth of the particular division of labour established in the electric power industry in Ontario – HEPC generation and transmission coupled with municipal power distribution – the myth drew its strength from pragmatic institutionalizations of ambivalent ownership. In this manner, the HEPC reflected, in Tuohy's words, 'the Canadian aversion for either-or choices [and] the tendency to find distinctive ways of reconciling divergent concepts.' All three major traits of institutionalized ambivalence she enumerates held featured roles in the history of the commission. Generally, the

HEPC legitimized competing principles with regards to its ownership through legislative statutes, while executive appointments allowed these principles to co-exist to the satisfaction of provincial and municipal leaders in a process which favoured elite accommodation.[2] In this manner, statutes and appointments were used in such a way that a choice need never be made over whether the HEPC, although created by the province, was a provincial government enterprise or the trustee of a municipal cooperative. When a choice was forced in 1973, the competing principles survived the establishment of Ontario Hydro, but only in their statutory form.

Statutes and appointments outlined the HEPC's character and its relations with both the government and the Ontario Municipal Electric Association (OMEA), the representative of the municipally owned local distribution commissions. What distinguishes these two instruments is that statutes are a means for formal control and appointments are a channel of informal influence. The government, in the HEPC's early years, used the formal control of statutes to disguise the actual extent of its involvement in the electric power industry. By melding government ownership with some of the competing principles of municipal cooperative ownership, statutory enactments ensured municipal support for the initiative and simultaneously defused the private sector's complaint of government ownership. Subsequently, appointments were used to institutionalize the divergent principles further through elite accommodation. Beginning with those made by Premier Howard Ferguson in 1925, the OMEA received representation on the commission symbolically equal to what the government received by statute, according both equal treatment as the main constituencies in the HEPC's decision-making. The Ferguson formula for appointments, by giving legitimacy to the municipal cooperative view, also served to disguise the extent of government ownership by governments that were essentially committed to private enterprise.

The lasting significance of the principle of balanced government and OMEA representation was such that it would reach, utilizing Andrew Heard's classifications, the status of a 'semi-convention.' As such, the formula was less than a 'fundamental' convention – one essential to the operation of the political system – but more than a 'usage' convention – one derived from a pattern of behaviour not based on a 'reason or principle.' Although the Ferguson formula for appointments might be 'occasionally disregarded without significant impact,' it provided a compelling reason for respecting its content as a general rule.[3] By adhering to the powerful myth that the HEPC was the trustee of a municipal cooperative in its use of appointments, the government avoided the wrath of the OMEA, a concern it took especially seriously during elections. With appointments being more vulnerable than statutes, the symbolic basis for the cooperative contention did not survive the creation of Ontario Hydro. The semi-convention could be broken at this time because the appointment formula had become an anachronism with the plethora of interests the utility had come to affect.

Despite the institutionalization of ambivalent ownership through statutes and appointments, the HEPC was clearly intended to serve government purposes. As a potential lever of economic growth and industrial development in the province, the creation of the commission was part and parcel of the phenomenon of 'province-building' as it relates to turn-of-the-century Ontario. Although the concept of province-building has been utilized to explain the postwar growth of provincial government activities vis-à-vis the federal government, particularly those which arose from the advent of the welfare state,[4] it has had a broader interpretation. In the larger context of defensive-expansionism – the phenomenon which historically has shaped Canada in the shadow of the United States – the phrase has been used to describe how provincial constitutional jurisdiction over natural resources has been the basis for provincial economic development strategies.[5] In this view, province-building has always co-existed in Canadian politics and public policy in a dynamic relationship with 'country-building.'[6] The HEPC was clearly intended to serve province-building purposes from its inception and Ontario Hydro continued this tradition until a rethinking of the corporation's mission occurred in the 1990s.

Marsha Chandler's analysis of the degrees of intrusiveness of crown corporations in the private economy can be used to track the HEPC's and Ontario Hydro's role as an agent of province-building. In her comparative study of provincial crown corporations, she identifies three types of intervention on the basis of the purposes served by each. All three are relevant to the utility's history. Facilitative corporations, the most predominant type, supplement and support private sector economic development and are not viewed as an extension of state control. This stems from their minimal infringement on the autonomy of the private sector. Redistributive corporations, for their part, challenge the distribution of economic and political benefits and thereby extend state control over the economy. Nationalistic corporations are an amalgam of facilitative and redistributive corporations in their enhancement of the provincial interest at the expense of other jurisdictions.[7] With the HEPC and Ontario Hydro having been assigned all three purposes separately and simultaneously, depending on the circumstance, they provide important markers of the evolving policy expectations of the government.

In assessing the impact of the province-building mission on the HEPC's and Ontario Hydro's character, the analysis of Jean Laux and Maureen Molot provides another comparative reference point, that of international political economy. By outlining how the objectives of crown corporations have changed over time, they establish a generational context in which to appraise the utility's development. First-generation crown corporations are those which nationalized whole industrial sectors, and usually those in financial trouble. Playing a passive role in the economy, they 'in effect service private enterprise at public expense.' Although the first-generation crown corporations examined by Laux and Molot appeared four decades after the HEPC, this category essentially describes the stage to which it had evolved by the mid-1920s, except that it was financially self-sufficient. Further changes to the character of the commission by the late 1960s make it comparable to what Laux and Molot categorize as second-generation crown corporations. Rather than being passive, such corporations have taken a directive role in the economy to stimulate industrial growth and to 'avoid social dislocation and eventual economic slump.' With capital intensive industries as the chosen vehicle, one example relevant to the HEPC and Ontario Hydro was nuclear power. Third-generation crown corporations arrived in the 1980s in internationally competitive and profitable industries, with their objectives being to 'counteract vulnerability to external change' in the new global economy. By representing a switch in the orientation of the state to business logic from social purpose, they differ from their predecessors in that they require low visibility and greater flexibility from government policy.[8] Ontario Hydro now struggles with this transformation.

Throughout their histories, the HEPC and Ontario Hydro have served the varying purposes and objectives outlined by Chandler and Laux and Molot. These two analyses will be used alongside Tuohy's notion of institutionalized ambivalence and my own focus on statutes and appointments as the analytical framework of the book. Following the introductory overview below, each of the five main eras of the HEPC's history constitute one of the book's substantive chapters, with the history of Ontario Hydro forming an epilogue that continues the HEPC themes to the present.

The ambivalence of the HEPC's ownership began with its unique creation as a hybrid of organizational forms. For what were essentially political reasons, it was given crown corporation decision-making, municipal cooperative schemes of expansion and debt retirement, and government department financing and cabinet representation, all alongside a vibrant private sector. By establishing a hybrid, the Ontario government served its interests in a pragmatic fashion. The outcome launched the government into the electric industry but, by obviating the need to nationalize private companies, left the government without responsibility for a first-generation crown corporation or public ownership monopoly. The government nonetheless crafted important but not obvious powers to control the HEPC by weighting the hybrid to a government department. The lesser corporation and cooperative elements, through their visibility, served to distance the government from political responsibility. Given the political implications of government ownership, this institutionalized ambivalence was considered the most suitable organizational form for pursuing the government's facilitative, redistributive, and nationalistic economic objectives.

While the HEPC had been created in co-existence with a vibrant private sector, Adam Beck took it far beyond its original mandate simply to operate a transmission company that supplied outlying municipalities. His twenty-year chairmanship of the commission was stamped indelibly by a personal leadership that fended off challenges from all quarters. Beck extracted corporate autonomy for the HEPC at the expense of the government's departmental controls, and bolstered the notion that it was the trustee of a municipal cooperative rather than a government enterprise to aid in this process. And against the wishes of his own government, he entered into direct competition with the private sector, successfully driving it out of the electric power industry. By establishing a virtual public ownership monopoly, Beck's redistributive project propelled the HEPC into a first-generation crown corporation broadly facilitative of economic growth.

Rather than resolve the HEPC's ownership after Beck's death in 1925, Premier Howard Ferguson institutionalized the ambivalence further through the government's power to appoint the commissioners. Alongside a non-partisan chairman from outside the government and the commission, the Ferguson formula for appointments granted the OMEA constituency representation on the HEPC symbolically equal to what the government received by statute. Besides submerging the previously disparate government and OMEA views on the HEPC's ownership, the Ferguson formula served to unite the government and the association in opposition to federal jurisdictional claims on the St Lawrence and Ottawa rivers. To avert the federal government's threat to its status as a first-generation crown corporation, the HEPC adopted a nationalistic stance. However, the soundness of buying Quebec power was undercut by the electricity surpluses recorded during the depression, leading to a second nationalistic policy to break the Quebec contracts. Given the demands of war production, the HEPC returned to a primarily facilitative role, but had nonetheless secured Beck's legacy of public ownership monopoly. By the 1943 election, the Ferguson formula was raised to the status of a semi-convention.

In the early postwar era the HEPC's facilitative crown corporation role emerges as a crucial instrument for promoting industrial growth. Given that this economic environment occurred in a setting of one-party dominance, the government wanted the commission to operate with policy sensitivity for its economic objectives. This development elevated to a new height the ongoing dispute over whether the commission was a government corporation or the trustee of a municipal cooperative. Although the scope of the HEPC had moved beyond the cooperative notion, the government and the OMEA engaged in a pitched battle – with the association protecting its representation on the commission and the government seeking to undermine the same through statutory changes. The renewal of the ownership dispute only reached a respite with the departure of the last minister to serve as a commissioner in 1963. Although the resilience of the ambivalent ownership triumphed, this outcome nonetheless left the impression of an OMEA victory.

While the OMEA could be happy that its municipal cooperative contention had survived, the institutionalized ambivalence of the HEPC would be persistently challenged thereafter. The commission's new role as the lead actor in a government-supported industrial strategy based on nuclear power drove previously unsuccessful reform. A shared victory would again factor into the outcome, but this time it was one that favoured the government. For this reason, the formal re-creation of the HEPC as Ontario Hydro in 1973 was of great significance. It represented the first major revamping in the commission's sixty-seven-year existence. In the face of the government's commitment to reform, the OMEA was unable to provide a viable alternative to the government's vision of a modern crown corporation. Premier Bill Davis's government succeeded in undermining the symbolic appointment basis of the municipal cooperative contention by creating Ontario Hydro with a large board of directors. The protests of municipal leaders, however, ensured that the statutory basis for their contention largely survived unamended.

By giving it a modern corporate structure, Ontario Hydro was set on the path of completing the HEPC's transformation to a second-generation and nationalistic crown corporation. The stimulus for this revamping was the utility's lead role in a nuclear industrial strategy. Within ten years this scheme would collapse in the face of recession and conservation, causing Hydro to search again for new purpose. Having never failed to find purpose in planning supply in the expectation that demand would catch up, Hydro wallowed through the 1980s experimenting with demand management and conservation. When an even more fundamental economic dislocation in the early 1990s made clear that further expansion was not an option, Hydro sought to reform itself in the image of a third-generation crown corporation. What now stands in the way of the flexibility it desires to have a more commercial orientation has been its visibility as a statutory creation and public monopoly. In this new environment there have been calls for privatization of Hydro, whether in full or in part, but the institutionalized ambivalence of its ownership complicates such an outcome.

2

Turn-of-the-Century State Intervention: Creating the Hydro-Electric Power Commission, 1902–6

The origin of the Hydro-Electric Power Commission was historically significant because it began as an organizational form very different from other early examples of public enterprise. It was thrust into an awkward, ill-defined co-existence with the private sector, and its initial character displayed a confusing degree of autonomy from government. Although it was created by government, no private companies were nationalized for the HEPC's establishment in 1906, and no emphasis was placed on its operation being or becoming sector-wide. Its relationship with the government also resembled a crown corporation only in part. Given that it began its mission at the behest of municipal leaders and under the direction of a government minister, in many respects it had the appearance of a municipal undertaking and a government department. For these reasons, the HEPC differed markedly from the typology of a first-generation crown corporation, although this was what it later came to represent.

The most striking feature of the anomalous nature of the HEPC's initial character was the institutionalized ambivalence of its public ownership. The key to uncovering the precise components of the outcome lies in tracking the alternatives which the government might have followed at the time of the commission's creation. There were four possibilities: increased regulation of private enterprise; promotion of cooperative municipal ownership; creation of a department of government; and creation of a crown corporation. In the light of these alternatives, the government constructed a pragmatic, politically rational response.[1] The HEPC's features reveal that the government, rather than creating the commission in the image of one of these alternatives, established a hybrid of a government department, crown corporation, and municipal cooperative that would co-exist with the existing private companies in an environment where neither was subject to regulation.

The relative importance of the elements of each alternative in the HEPC's initial hybrid structure lay in the weighting of the government's means of formal, statutory control and informal, political influence. While on the surface these means reveal that the HEPC received the outward character of a crown corporation with ostensible autonomy from government and some features of the trustee of municipal cooperative to ensure municipal support, this was only part of the story. The government gave the strongest weighting to the less obvious but powerful controls that characterize a government department. This mixture demonstrates that the purposes underlying public ownership were of critical importance to the government. The HEPC's hybrid structure was perceived as the best means to facilitate economic development, redistribute the inequities of monopolization, and produce nationalistic advantages for the province.

Niagara in Politics

Before the HEPC's creation in 1906, the ways in which economic development could best be promoted by electric power in Ontario were a matter of great political debate. The controversy arose at this time because technological advance had made it possible to transmit electric power over relatively long distances. This advance, although of international significance, attracted considerable attention in Ontario because of the province's immense water power resources, particularly at Niagara. Since they did not have an indigenous supply of coal to generate power, Ontarians were naturally captivated by water power's capacity to deliver the second industrial revolution occurring elsewhere.[2] The potential of water power to generate inexpensive electricity and reduce dependence on American coal, which was sensitive to price escalations and labour unrest, made it a valuable resource for the province. In this setting, the existing minimal regulation of private enterprise in the industry was under siege.

Advances in transmission technology sparked two interrelated controversies which in turn stimulated the search for an alternative to private sector regulation. Both stemmed from the fact that the Electrical Development Company (EDC) was seeking a franchise to develop water power on the Canadian side of Niagara. The company was a syndicate formed in 1902 to complete the vertical integration of related companies. Its specific role was to generate electric power for transmission by the Toronto and Niagara Power Company, distribution by the Toronto Electric Light Company, and consumption by the Toronto Street Railway Company. Moreover, the franchise it sought was the last of three made available, with the other two already held by American-owned companies that exported their power to nearby Buffalo. Thus, one controversy over Niagara centred on the merits of a vertically integrated monopoly serving Toronto. The other centred on the likelihood that other communities of southwestern Ontario would not be served by Niagara power, leaving them to rely on power generated by coal. The two controversies converged to form a united search for an alternative to the private character of electric power development at Niagara.

The debate over the concentration of the electric power industry in Toronto was not unique among utilities in North America. In both Canada and the United States the monopolization of utilities was countered by 'civic populism,' a movement which sought to reverse the trend through 'local control and municipal freedom.' In Canada, however, the movement was unique in two regards: almost all its leaders were elected officials; and in Ontario, the centre of the dispute, it had a very broad base. This latter point was significant. Support from local boards of trade and the Canadian Manufacturers' Association meant the business community was not united in defence of the property rights of the private monopolists, and support from both management and trade unionists indicated the movement had 'crossed the class divide.'[3]

Even prior to the progress in transmission technology, Toronto had been a centre for civic populism. The city council had applied many times to the provincial government for approval to build its own coal-generated power plant. The objective was to sell electric power at a lower rate than that offered by the profit-seeking Toronto Electric Light Company, but permission was never granted because of the competition it posed. Moreover, the council's repeated applications were made in vain after 1899 because of an amendment to the Municipal Act initiated by James Conmee, Liberal MPP for Algoma West. Known as the Conmee Clause, it extended an existing prohibition on municipalities competing with private water utilities to include gas and electric utilities. For a municipality to enter any of these fields, the act required it to purchase the existing company, although subject to price arbitration.[4]

The second dimension of the controversy over Niagara, the prospect that southwestern Ontario communities would be cut off from its benefits, was particularly worrisome to the business and municipal leaders of that region. Without an injection of inexpensive electric power, they feared economic decline. At a meeting of the Waterloo Board of Trade on 11 February 1902, E.W.B. Snider, a former Liberal MPP and leading local industrialist, first raised this gloomy scenario. In a challenge to the boards of trade of the region, he called for the formation of a power cooperative of municipalities and manufacturers. With this historic meeting, municipal cooperative ownership emerged as an alternative to the existing regulation of private enterprise.

Given the interest aroused in cheap power, Snider and W.B. Detweiler, a former president of the Berlin (now Kitchener) Board of Trade, organized a public meeting in Berlin on 9 June 1902. The possibility of servicing the region with power from Niagara was the focus of attention. To this end, C.H. Mitchell, the invited guest speaker who was the engineer of the American-owned Ontario Power Company, informed the meeting of developments in the industry. Alderman F.S. Spence of Toronto, who had initiated Toronto's applications for a municipal power plant, was also invited to relate the trials of Toronto city council.

At the meeting Snider reduced the municipal leaders' options to three: the private power companies at Niagara could build transmission lines to the municipalities; the provincial government could get into the business of power transmission; or the municipalities could form their own power cooperative. The meeting adopted a motion to create a municipal cooperative and selected a working committee to pursue this end. At the committee's first meeting on 30 June it was decided that Snider, along with Alderman Spence and Alderman J.H. MacMechan of London, should speak to Liberal premier George Ross to outline their plan. Detweiler, for his part, set out to assemble figures on electric power demand for each municipality interested in the cooperative scheme and the bulk prices charged by the companies generating power at Niagara.

The municipal leaders had to move swiftly as they were competing with the EDC for the remaining franchise at Niagara and the EDC had already made its application to the government. Even though the franchise had yet to be awarded, Snider reported back to the southwestern Ontario municipal leaders on 20 October 1902 that his delegation's meeting with Premier Ross had not proved fruitful. Ross was neither willing to reserve water rights for a municipal cooperative nor prepared to have the government operate a transmission company. He felt justified in his stance because he was unwilling to incur provincial debt for only one region of the province and would not accept the municipal leaders' contention that, in H.V. Nelles's words, the 'public interest' so closely paralleled their 'manufacturers' interests.' Moreover, Ross could not understand their fears of monopoly, suggesting the number of companies operating at Niagara was evidence of competition.[5]

The stakes were high for future development on the Canadian side of Niagara. In awarding the franchise, the Ross government had to choose between the well organized and financed EDC syndicate, formed with the intention of supplying Toronto, and the embryonically organized and politically disaffected southwestern Ontario business and municipal leaders. The choice was made less difficult by the fact that the cooperative scheme had yet to be shown to be economically viable and would likely require government financial assistance. More germane to the outcome, in W.R. Plewman's view, was the fact that 'the electric companies had vigilant and resourceful friends [at Queen's Park] who repeatedly checked and baffled the forces campaigning for cheap power.' Indeed, Ross, while premier, was president of the Manufacturers' Life Insurance Company and the EDC principals also served as company officers and directors. In 1906, moreover, it would be revealed that the company had investments in the EDC.[6] Thus, the municipal leaders were likely not surprised when the EDC was granted the franchise by the Ross government on 27 January 1903, but they remained undaunted.

Notwithstanding this defeat, the municipal leaders had become a pressure group of sizeable political force that would have to be heard in the future. By the time of the second meeting at Berlin on 17 February interest in the municipal cooperative had expanded to attract sixty-seven business and municipal leaders, including representatives from nineteen councils. At the meeting Snider and Detweiler sought approval for a plan to create a transmission company cooperatively owned by municipally owned distribution commissions. The advantage of this scheme was that it avoided the high capital cost of generation because the transmission company would purchase power from private companies at Niagara. Recognizing that their plan required legislation, they recommended that the government permit a municipality to join the cooperative through a vote of its electors. However, the motion to adopt the Snider-Detweiler plan was amended with bipartisan support on the initiative of Adam Beck, the mayor and Conservative MPP of London, and Thomas Urquhart, the Liberal mayor of Toronto. Their amendment, a trailer to the main motion, called on the municipal leaders to 'urge upon the Ontario Government the advisability of government building and operating [the transmission system] as a government work.'[7]

The amendment was of great significance. With it, government ownership had emerged as an alternative to municipal cooperative ownership, dividing the municipal movement over its official objective. Ten days later, on 27 February, an embarrassed Snider, the representative of the municipalities, had to demand that Premier Ross proceed with one of the two alternatives, after having previously made known his own preference for a cooperative. Given the choice, Ross ruled out government ownership and accepted municipal cooperation, to which he did not object in principle. The government passed an act, dubbed the Ross Power Act, with this intent on 12 June 1903.[8] However, the act did not bring an end to calls for government ownership.

The terms of the Ross Power Act suggest that it was designed simply to prescribe the rules regulating municipal entry into the electric power industry and not to lend government endorsement to municipal cooperative ownership. Rather than encourage municipal ownership, its chief purpose, James Mavor reports, was to 'prevent the municipalities from plunging in unrestricted fashion into reckless adventure.' The act established no means for formal, statutory government control over a cooperative through an active or even passive role for the cabinet, nor any provision for informal political influence through government appointments to their boards. Moreover, the act, by not committing the government in any way to underwrite the cost of such cooperatives, meant that they were to be treated as strictly private concerns. This explicit avoidance of political and financial responsibility was most evident in the act's placement of supervisory responsibility with the chief justice of Ontario rather than cabinet.[9] In a peculiar twist of political responsibility, the chief justice at this time was William Meredith, the former leader of the Ontario Conservative party.

The act set out in two distinct stages the details of how municipalities were to proceed with the establishment and operation of power cooperatives. In the first stage, councils were required jointly to appoint an inquiry commission of two to four persons and one engineer. Its task was to produce a detailed report on the cost of the works and the price of electric service for every interested municipality. This report would then serve as the basis for individual municipal councils to review their participation in the proposed cooperative. If a council favoured joining the cooperative, it first had to publish the plan, including the total cost of the works, and then hold a bylaw referendum under the provisions of the Municipal Act. With this stage completed, no further role existed for the inquiry commissioners.[10]

The second stage of the Ross Power Act outlined the procedure for establishing a permanent municipal power cooperative. Following the ratification of the municipal bylaws, the chief justice was to appoint 'at pleasure' a board of commissioners from persons nominated by the municipalities. Municipal councillors and officers were specifically disqualified from assuming these operational control positions to ensure a businesslike operation. The chief justice was also responsible for appointing the cooperative's auditor, but the audited annual report was directed to neither the chief justice nor the cabinet. It was to be submitted only to the commissioners and the secretary of the Bureau of Industry, like the annual report of any private company, and copies were to be made available to the participating municipalities, presumably like any company shareholders. The commissioners, for their part, were assigned responsibility for the cooperative's day-to-day business and for ensuring that it was self-sustaining. All power contracts with third parties, however, fell under the chief justice's supervisory responsibility, as did the adjudicative function of arbitrating disputes with third parties.[11]

The Ross Power Act went to great lengths to ensure that the provincial government would have no financial responsibility for municipal cooperatives, but its restrictions made the financing rules convoluted. The process was to begin with the commissioners' selection of a trustee to act as a third party financial agent. The trustee was to receive from each municipality the debentures approved in their bylaws and then transfer these funds to the commissioners for construction purposes. In exchange, the commissioners were to transfer to the trustee bonds that were equal in value to the debentures and backed by a non-preference lien on assets. The trustee was also required to return to each municipality amounts sufficient to maintain a sinking fund for its debentures. In cases where the debentures issued were not sufficient to complete the works, the chief justice could authorize the issue of special bonds with a first lien on assets. The lone financial concession was that municipal power cooperatives would be exempt from taxation.[12]

In general, the Ross Power Act permitted the municipalities to engage in any aspect of the electric industry, including generation at Niagara, but provided little assistance for them to be successful. Its overarching theme was to insulate government from responsibility for any municipal power cooperative which might result. It did so by assigning a supervisory role to the chief justice and by removing the cooperatives from municipal politics. The most significant problem with the act from the municipalities' standpoint, however, was that it did not rescind the Municipal Act's prohibition on competing with private municipal utilities (the Conmee Clause) or amend the Municipal Act's prescribed limits on the total indebtedness of municipalities.

Competing Commissions of Inquiry

The municipalities interested in proceeding under the terms of the Ross Power Act met in Toronto on 12 August 1903, two months after the passage of the act. Although sixteen municipalities were represented, only seven participated in the establishment of the Ontario Power Commission, the first and inquiry stage outlined in the act. They appointed Snider, a former Liberal MPP, as chairman and Adam Beck, a Conservative MPP, W.F. Cockshutt, a prominent manufacturer who became a federal Conservative MP in 1904, and P.W. Ellis, president of the Canadian Manufacturers' Association and a Conservative, as commissioners. The commissioners then selected Professor R.A. Fesserden of Washington, DC, as its engineer and fifth commissioner. Detweiler was not among the members because he had left the municipal committee to establish the privately owned Algoma Power Company in 1902.[13] Although nominally bipartisan, the open political affiliations of the commissioners and the preponderance of Conservatives had the potential to keep the Snider Commission's work mired in partisanship and bring embarrassment to the Liberal government.

Government ownership was the issue likely to cause both the Snider Commission and the government problems. Although the municipal movement had been consolidated into a government-sanctioned fact-finding inquiry for a cooperative, it had never officially rejected government ownership as an alternative. Government ownership had first been adopted on Beck's initiative and would in fact resurface at his urging. Despite being a Snider commissioner, Beck would capture the attention of municipal leaders and the Conservative party with government ownership long before the Snider Commission submitted its report. Although the report would have to limit itself to recommendations on a municipal cooperative, Beck's effort would not be for naught. The report would also be useful for the cause of government ownership.

By the time of the Snider Commission's first meeting in August 1903, Beck had vaulted to a leadership position in the municipal movement rivalling that of Snider. This was surprising because he had only been elected a mayor and MPP in 1902 and had attended his first municipal meeting in February 1903. Both the politics of the period and Beck's political style played a significant role in his rise. In the 1902 general election the Ross Liberals had barely clung to power with fifty of ninety-eight seats and showed little sign of being able to recover in time for the next election. This meant, in Nelles's view, that Snider's influence suffered because much of it stemmed from his 'close relationship' with Premier Ross. Beck, on the other hand, was able to fill this void because he was seen to have influence in the prospective government. In addition, the Conservatives were generally the party of urban Ontario in this era, both municipally and provincially, a consideration which put Beck in better stead with fellow municipal leaders than Snider.[14] With this potential for rivalry between Snider and Beck threatening to fracture the tenuous consensus for a municipal cooperative, the Snider Commission proceeded with its business.

Beck's impetuous political style stood in strong contrast to Snider's quiet diplomacy. In him, Nelles writes, the 'movement acquired what it had long needed – assertive leadership unrestrained by modesty or self-consciousness.'[15] This left Beck better suited than Snider to exploit the general anti-monopoly sentiment then existing in Ontario and the specific political opportunity that civic populism presented in the politics of Niagara. It mattered little that he was responsible for keeping the municipal movement and the Snider Commission divided on its preference for either municipal cooperative ownership or government ownership.

Although Snider remained the nominal leader of the municipal movement as the chairman of the Ontario Power Commission, Beck came to eclipse him as its leading figure. In the prelude to a 1905 provincial election, Beck had successfully rallied public sentiment for cheap power, and had persuaded his Conservative colleagues to adopt the popular cause as party policy. This was especially significant as the Ross Liberals were known to have made it difficult for the municipalities to get their cooperative operating. Moreover, since Beck was only forty-six in 1903, compared to Snider who was sixty-one, the municipalities had gained a leader whose age, character, and political affiliation demonstrated that he could lead their movement through to the realization of its broad objective of cheap power.

With the 1905 election looming, the Ross government sought to bolster its electoral prospects through the cheap power issue as well. Aware that the Conservatives were leaning towards government ownership, it acted to secure power for a municipal cooperative. The EDC, in the knowledge that the government was in danger of defeat, had applied to the Queen Victoria Niagara Falls Park Commission, the government regulatory body, for an additional 100,000 horsepower of water rights. The Park Commission, undoubtedly aware of the politics involved, deferred the request to the government as a 'matter of public policy' on 11 November 1904. The government, in turn, used this opportunity to insist that the EDC allocate one-half of the new concession for a municipal cooperative. When the company balked, its application was overtaken by the onset of the election. Anticipating a Liberal loss, the EDC eventually accepted the terms on 9 January 1905, during the campaign.[16]

The prospects for government action on cheap power were greatly enhanced when the Conservatives won the provincial election in January 1905 with sixty-nine of ninety-eight seats after thirty-four years of Liberal governments. Electric power policy had been one of the few major political issues in the years leading up to the election and a key issue which differentiated the parties and their leaders in the campaign. Premier Ross had defended his policy of promoting corporate competition at Niagara as the means to foster cheap power. Conservative leader James Whitney, for his part, condemned the Liberals for permitting the industry to develop for the export market and for favouring the EDC over the aspirations of the municipal leaders. Since Ross had corporate connections to the EDC principals through the Manufacturers' Life Insurance Company, this was good politics.

By 1905 sentiment in the province had largely swung to favour government ownership over municipal ownership, while the rhetoric of municipal leaders had similarly swung to public power from cheap power. The two factors which made this election such a watershed were the same ones that now made the objective of public power obtainable. First, turn-of-the-century Ontario was experiencing a large rural-to-urban demographic shift that undermined the longstanding rural base of Liberal support and enhanced the urban base of Conservative support. Thus, electric power, as a largely urban issue, played a key role in determining the election outcome. Secondly, advances in technology, according to J.E. Hodgetts, provided the stimulus for government to switch from an arm's length to a hands-on administrative mode at this time. In addition, the Liberals had greater difficulty than the Conservatives in responding to the implications of the demographic shift and technological change on the power issue. As Charles Humphries explains, the 'Liberals, so long in power, [had] developed an inclination towards the view of the power producers, while the Conservatives, in opposition, understood the problem from the consumer point of view.'[17]

Not surprisingly, when the Whitney government took office on 8 February, it enjoyed the support and goodwill of the public power movement. Whitney responded by taking three decisive actions favourable to the movement. He made Beck a minister without portfolio and designated him the government's principal spokesman on power issues. Then in May he rescinded the agreement the Liberals had made with the EDC during the election to grant additional water rights, a measure which the company was now pressuring him to fulfil. Although the Conservatives had charged that there was politics in the agreement, their pretext for not respecting it was that it had not been formally ratified by the legislature. The action was of great significance because it halted any new private development at Niagara.[18] Whitney also addressed the pressing issue of the suitability of private development governing the industry at Niagara and elsewhere. On 5 July he established the Hydro-Electric Commission of Inquiry, giving it much broader terms of reference than those of the Snider Commission. The scope of the inquiry included obtaining financial and technical information from the private companies and the scale extended to the whole province, not just Niagara. Whitney appointed Beck as chairman and P.W. Ellis, the Snider commissioner, and George Pattison, a Conservative MPP, as commissioners.[19]

All three of Whitney's initiatives were politically astute. Giving Beck responsibility for both power issues and the inquiry ensured municipal support for the new government. Halting future development at Niagara until the inquiry reported bought time for the government to establish its own policy on regulating private development. And establishing the inquiry permitted the government to explore the alternatives of either assisting the formation of the municipal cooperative or assuming responsibility for transmission to municipalities itself. Moreover, the fact that a commission of inquiry could be created by the cabinet alone, avoiding the legislature, was an important consideration.[20]

Hydro-electric issues remained prominent throughout the winter of 1906 with the impending release of reports from the Snider and Beck commissions. The struggle between Snider and Beck for the direction of the municipal movement, especially with Beck remaining a member of Snider's commission, was highlighted when Ellis, also a member of both commissions, resigned from only Beck's commission on 23 January. The pretext was ill-health, but it probably resulted from Beck's mistreatment of Snider. This was a temporary set-back to the Beck Commission because the powers of its commissioners were joint, meaning that with Ellis's resignation, it could not continue. The commission was reconstituted on 26 January, with John Milne of Hamilton filling the vacancy.[21] Nonetheless, this was not the end of the tension between Beck and Snider.

The reports of the Snider and Beck commissions were released within a week of each other. Snider's report of 28 March was moderate, as expected, adhering to the original scheme for a municipally-owned cooperative and advocating no provincial government involvement. The harsh criticism of the character of private development given in Beck's report of 4 April did suggest strong action could be expected from government, but it made no recommendation for either a municipal cooperative or provincial government ownership. In public statements, however, Beck made his position quite clear. He was of the opinion, according to Nelles, that 'since the province would have to act as banker and coordinator of the municipal cooperative ... it might as well take an active role in the running of whatever system might be decided on.'[22] Given his advocacy of government financing and leadership, Beck had the municipalities on his side even though Snider remained closer to the original sentiment for municipal ownership.

Recognizing that his vision of government transmitting power to the municipalities would be a departure from its historic functions, Beck, ever the skilful public figure, set in motion a campaign to promote his view and undercut both Snider's profile and the Snider Commission's recommendations. His efforts included helping revive the municipal movement, left dormant after being subsumed into the commission for the purposes of the Ross Power Act. The movement was reorganized as the Municipal Power Union on 23 March 1906, two weeks prior to the release of the Beck report, and it willingly agitated for his objectives. In addition, Beck openly appropriated the Snider Commission's research to keep the municipalities focused on his own agenda, politically sidelining Snider in the process. According to Plewman, Snider was astounded when he discovered that the figures supporting his recommendations, worked out by consulting engineers at the expense of the municipalities, were 'published in newspapers and credited to Adam Beck.'[23]

Although the terms of the Ross Power Act officially directed the Snider report to the municipalities that had participated in its formation, the document spoke to the much broader audience of all Ontario municipalities, the municipal movement's private sector detractors, and the provincial government. In essence, the report recommended that the municipalities proceed with a cooperative for both transmission and generation of electricity. And, although the report did not explain why the cooperative should assume the high capital cost of generating power, it did not conclude, as Nelles writes, that the municipalities could not afford to build a generating station.[24] The report was convinced that the cooperative, if it was operated as a 'business institution and absolutely divorced from politics,' would be successful. This success would be enhanced, the report argued, if the cooperative expanded to embrace eleven adjacent municipalities in addition to the seven which had initiated the inquiry. The report worked out the costs and prices of supplying electricity to both groups, noting the 'advantageous conditions' of the enlarged cooperative. Indeed, the report went on to note that if the cooperative became province-wide the 'aggregate savings would reach a colossal scale.'[25]

The best indication that the Snider report was directed to a broad audience was the great weight it placed on a municipal cooperative's capacity to promote economic development and produce a comparative advantage for the province. The seven municipalities which initiated the study had already been convinced of these benefits. After noting the lower rates for each category of ratepayer, the report explained how these rates would facilitate 'productive and competitive efficiency' in manufacturing. This, in turn, was claimed to be able to attract the 'enterprise of others.' As an indication of the size of this impact, the Snider Commission's engineer calculated that Niagara could generate more power than the total then generated with coal in the United Kingdom. This electric capacity, he felt, was 'destined to make [Niagara's distribution area] perhaps the foremost manufacturing centre in the world.' In sum, the Snider report argued that a vibrant municipal cooperative would exercise an extraordinary influence upon the future of Ontario.[26]

The Snider report stressed the economic benefits of a municipal cooperative to counter and refute criticism of municipal ownership. Without seeking a direct confrontation with private companies, the main protagonists, it stated only that their opposition was 'coloured by considerations of self-interest.' In response to those opposed on principle, the report argued that municipal ownership simply 'represented the voluntary efforts of society to work out a more efficient civilization.' And despite the report's expansionist tenor regarding the proposed cooperative, it noted that 'ample field' would remain for private companies. By taking this non-confrontational position, the report did not then go on to recommend, as Nelles contends, that a permanent commission 'regulate the entire hydro-electric industry.' On the related matter of whether the electric power industry was too much of a risk for municipal capital, the Snider Commission's engineer stated that there was no 'engineering objection.' The reason, in his professional opinion, was that water power developments would not be rendered 'obsolete through invention.'[27]

The Snider report also addressed the hurdles to be overcome in order to get the municipal cooperative operating. Not surprisingly, the report focused on the Conmee Clause of the Municipal Act and the restrictive financing provisions of the Ross Power Act. On the former, the Snider report noted that three of seven municipalities in the study had private utilities that would have to be bought out, and that the utilities in two, Toronto and London, would refuse to sell. The report, however, only stated that the Conmee Clause would 'interfere' with the plan outlined in the report. It did not go on to recommend, as Nelles maintains, that the clause be repealed. Since repeal meant there would be competition without compensation, Snider for one would not likely have been onside because he owned the electric properties in Waterloo and had investments in Detweiler's Algoma Power Company. Beck, for his part, may very well have accepted the word 'interfere' knowing he had the influence to see that it would be translated as a recommendation for repeal.[28] As for the financing provisions, the report recommended that the municipal debentures should be backed by a specific first charge on the construction bond issue, rather than a non-preference lien. This additional security was thought necessary to reduce borrowing costs until investor confidence could be established.[29]

The Beck Commission, which submitted only a partial report to the Whitney government, tackled the broader scope and scale of its mandate with some innovation. It divided the province into five geographically distinct water power districts and made Niagara the subject of its first report. This decision was expedient given Beck's desire to undermine the Snider Commission. With his own commission having been in existence for less than a year, he was able to utilize the Snider Commission's extensive research, and his report did not hide this fact. The Beck report, however, lacked the thoroughness displayed in the Snider report in terms of detailing costs, for example, and was short on specific recommendations. Mavor comments that its details were of the 'most meagre and unsatisfactory character.'[30]

The content of the Beck report would appear to suggest that it was produced for political purposes, specifically to belittle the Snider report's recommendation to construct a generating plant, to criticize the negative economic effects of private development, and to draw public attention to the poor stewardship of the Queen Victoria Niagara Falls Park Commission. Nowhere did the report recommend government ownership, although this has been the popular mythology. Despite its harsh criticism of existing government regulation and its lack of emphasis on municipal cooperative ownership, the Beck report did not propose, as Nelles writes, that the government create a 'provincial hydro-electric commission to regulate the affairs of the private companies and to distribute its own power to the municipal utilities.'[31]

On the issue of electric power demand, the Beck report virulently criticized the negative effects of private development to make its point on the facilitative potential of abundant, inexpensive water power. The report stated that demand existed for large quantities of power but insisted that this demand could only be unleashed if power was made available on an at-cost basis. It then condemned private corporations for working against the diffusion of electric power by not keeping their prices down. Citing Montreal, Buffalo, and Hamilton as examples, the report stated that

> the trend of affairs with private corporations ... has been, not to compete for business, ... but to amalgamate or otherwise destroy competition, and then to fix the prices according to the slight savings which they may be able to induce their customers to make. The natural result of this has been to force individual consumers, where the circumstances justified it, to install generating plants of their own, or to adhere to existing methods, rather than to place themselves at the mercy of large combinations formed for the purpose of preventing competition and keeping up the price of electrical power; and the same result occurs where there has never been a competing company.[32]

This critique of the constraints on electricity's economic potential under private enterprise was only a precursor to the report's attack on specific companies, particularly the Electrical Development Company.

The Beck report's criticism of private development was based on the deficiency of regulation as the instrument governing the industry. The minimal regulation then in existence was primarily restricted to monitoring the quantities and rates of power generated, which gave government little or no control over development. Under the terms of the franchise agreements between the private companies and the Queen Victoria Niagara Falls Park Commission, the companies were obliged to provide 'verified statements' of the power they generated and sold, and divulge the rates charged when requested by order-in-council. The Beck Commission sought to examine the statements only to discover that none had ever been submitted. On request, the information was voluntarily delivered directly to the Beck Commission by two small companies, but the larger EDC and the American-owned Ontario Power Company refused. The Beck report recommended that the statements be insisted upon and that the rates be requested by cabinet to provide a public record.[33] The report, however, did not make recommendations on how regulation might be improved.

Despite the intransigence of the large private companies, the Beck report was able to provide figures on the costs of generation and transmission of power because much of the work had been completed by the Snider Commission. In general, the report simply supplemented the Snider report's figures to arrive at others for the larger number of municipalities it determined could be supplied by Niagara. On generation in particular, demand scenarios in the Snider report's recommendation to build a plant were used to cost plants of three different horsepower levels at Niagara. The Beck report thereby circumvented the existing large private companies' refusal to divulge capital costs. Having provided generation figures, however, the engineering appendix of the report only recommended that a transmission company be established, which would purchase its power from private generating companies. The primary reasons given were the shorter construction time horizon and the greater flexibility permitted for demand to develop. No mention was made of who should own the transmission company.[34]

In sum, the Beck report did not propose to have the provincial government financially underwrite a municipal cooperative or own the transmission system. It did not discuss any requirement for formal, statutory government control over the transmission company or any need for informal, political means of government influence, such as appointments. The rhetoric in the report and Beck's own public statements did suggest strong action from government could be expected, but information on specifics would have to wait until legislation was introduced.

On the release of the Beck report, any common ground Snider and Beck might have previously shared on the municipal movement's objectives was now dispelled. By placing little emphasis on direct municipal cooperative ownership of transmission and by implying that government would assume a strong role, Beck's report and public statements undercut Snider, but Snider was too weak politically to restrain him. Premier Whitney was the only person who could take control of the power issue, but even he was outflanked. On 12 April 1906, eight days after the release of the Beck report, fifteen hundred 'public power' supporters marched on Queen's Park. According to Plewman, 'the voice of the people, under Adam Beck's expert coaching, was asserting itself in no uncertain manner.'[35]

This staged march on the provincial legislature had been preceded by a Municipal Power Union meeting held at Toronto City Hall. The assembled municipal leaders passed a motion calling for legislation to permit the government to appoint a permanent commission as the means of 'safeguarding the people's interests.' This commission would not only have broad powers to undertake 'generation, transmission and distribution' of electric power; it would also have the authority to regulate all electricity prices, including those charged by private companies.[36] Thus, the municipal movement had chosen to favour government ownership *and* increased private sector regulation. Desiring a body with the decision-making and autonomy of a crown corporation, the implication was that the municipal movement had abandoned cooperative ownership as its objective.

The Power Commission Act

On 7 May 1906 Beck introduced a government bill to create the Hydro-

Electric Power Commission as a permanent government body. Although he had been responsible for the broad outline of the bill, neither he nor any other member of the government was its actual author. The ghost author was Whitney's mentor and predecessor as Conservative party leader, William Meredith, the chief justice of Ontario. Meredith had maintained an active, although discreet, role in his party's affairs. Plewman, in fact, argues that he drafted much of the legislation, given that Beck was a 'political novice and not sure of himself.'[37] Meredith's involvement, however, was possibly of even greater significance to the outcome because he privately assured Whitney that the bill was a 'safe and conservative measure' and urged him to resist being 'stampeded into withdrawing or emasculating' the legislation. After significant amendment, the bill passed third reading three days later on 10 May, the last day of the session.[38] Despite the fact that Ross was still Liberal leader, it passed without division. His caucus voted with the government rather than support his objections to government involvement in the electric industry.[39] The features drafted into the Power Commission Act, as the HEPC's statute would later come to be known, made it dramatically different from the Ross Power Act, but the organizational form was less clear. In fact, it created the commission as a hybrid of a government department, crown corporation, and municipal cooperative.[40]

The official title of the Power Commission Act was 'An Act to provide for the Transmission of Electrical Power to Municipalities.' Although this title was non-committal on the ownership of the commission, it belied the dominant role the act granted the government in HEPC affairs. This role, readily evident in the placement of supervisory responsibility with the cabinet (rather than the chief justice) and the government's commitment to finance the undertaking, clearly differentiated the act from the Ross Power Act. It suggested, as Alexander Brady points out, that the commission conformed in broad essentials to a department of government. However, this role for government was not fully apparent, he adds, because the act stated that the HEPC was 'a body corporate.'[41] Although this did little to hide the fact that the government had appropriated the municipal initiative, the appropriation did satisfy municipal objections to the Ross Power Act, many of which were not even addressed in the Snider and Beck reports. Moreover, the appropriation accommodated, in part, the notion that the commission was the trustee of a municipal cooperative. This enabled municipal leaders to support the initiative willingly. The only actors not accommodated were the private companies, but they were appeased in two ways: there was no plan to nationalize them, and their objection to the HEPC regulating their prices was heeded.

In outward appearance, the HEPC had three features that suggested it was a crown corporation rather than a department of government or municipal cooperative. First, the commission's decision-making was made collegial through a board of three commissioners, rather than hierarchical through a minister and departmental portfolio. Secondly, the HEPC had the power to hire its own staff outside of civil service rules. And thirdly, the government, rather than the municipalities, was given the power to appoint all of the commissioners as well as designate the chairman. On its own, the power to appoint the commissioners indicated the commission was a crown corporation rather than a department because it provided the government with only informal political influence, and thereby was not in conflict with collegial decision-making. However, the act stated that one of the three commissioners must be a cabinet minister and two may be cabinet ministers. The legislation had actually been amended on 9 May 1906 to include this provision, having only provided that one minister may be appointed when first introduced.[42] Thus, the requirement that ministers sit as commissioners in effect formalized the government's otherwise informal influence in the HEPC, giving it a departmental character.

The provision for ministers sitting as commissioners was just one of three dimensions to the departmental character of the HEPC that differentiated it from a crown corporation. All three stemmed from the overarching role the Power Commission Act assigned to the cabinet in the commission's affairs. The second dimension rested upon the numerous statutory provisions for formal cabinet approval of the commission's business decisions. All of these provisions centred on the extensive powers of acquisition and expropriation granted to the HEPC, powers which were absent in the Ross Power Act and not included in the recommendations of either the Snider or Beck commissions. Under the Power Commission Act, the HEPC could request that cabinet authorize the acquisition or expropriation of any lands, works, or other property of private companies, including the power they generated. If cabinet gave its approval, the commission enjoyed all the powers found in the Public Works Act. The HEPC and its commissioners were also granted crown immunity, unless the attorney general of Ontario granted leave.[43] Under this umbrella of formal cabinet control and the privileges of a government department, the commission's form differed little from that of a department.

The financial provisions of the Power Commission Act were the third dimension to the HEPC's departmental character. The Whitney government addressed the financing deficiencies of the Ross Power Act by assuming the burden of underwriting the commission. In the process, it took control of all commission finances through three provisions. First, the Power Commission Act provided for the cabinet to raise capital for the HEPC in the name of the province rather than for the commission to issue its own securities either independently or with a government guarantee. This left the commission without a critical avenue of corporate freedom. Secondly, the act dictated that the HEPC's revenues and expenditures were to be managed by the government and audited by the provincial auditor, and it specified in particular that all revenues were to be paid over to the treasurer of Ontario to retire the commission's debt. This gave the HEPC the same financial status as a department, and ironically, meant that it was required to pay its employees, including commissioners, out of legislative appropriations, despite having the discretion to hire its own staff. Moreover, remuneration for commissioners other than those in cabinet (who received no extra remuneration) was also the prerogative of cabinet.[44] And thirdly, with these financial constraints, a formal annual financial report was neither necessary nor made a statutory requirement, leaving the HEPC without a visible marker of a corporate status. Thus, the Power Commission Act was framed as a trade-off. The HEPC received government financial assistance in exchange for government financial control which circumscribed its corporate autonomy.

Although the Power Commission Act gave the HEPC the appearance of a crown corporation and the character of a government department, it also accommodated the notion that the commission was a municipal cooperative of sorts on two fronts. One accommodation was the organizing principle that all HEPC expenditures were repayable by the municipally owned distribution commissions it supplied. The act instructed the HEPC to charge each participating municipality a proportionate share of (1) 4 per cent interest on the capital expenditures of the commission, (2) sinking fund payments to retire the securities issued by the government, and (3) general operating expenses of the commission. All of these municipal payments were to be paid over to the provincial treasurer, leaving the HEPC no role in administering the retirement of the debt.[45] Thus, the notion that the HEPC was the trustee of a municipal cooperative was enshrined in its statutory structure despite the fact that it was to be governed like a department.

The second accommodation of the notion that the HEPC was a municipal cooperative was the lead role the Power Commission Act assigned municipal councils in initiating the commission's growth. On this score, the act followed the early outline set by the municipal movement that was later incorporated in the Ross Power Act: the HEPC was only to supply power, and thus expand, after a municipality had held a bylaw referendum.[46] However, the act was more lenient than the Ross Power Act on a number of points. The bylaw approval was split into two stages, dividing the approval for a provisional contract with the HEPC for transmission of power from the raising of debentures for a municipal hydro commission's distribution works. Furthermore, the bylaw for the provisional contract did not have to specify a predetermined total cost to the municipality. Finally, the act exempted municipally owned distribution commissions receiving power from the HEPC from the strictures of the Conmee Clause.[47] This meant that Toronto and London, the major centres of electric demand affected, could establish utilities in competition with private companies.

Three trade-offs were made for the provisions favourable to the notion that the HEPC was the trustee of a municipal cooperative. The first was that the commission was given the power to regulate municipal rates and other matters, with appeals to be heard by the commissioners as well. The legislation, as first introduced, had also included the power to regulate private electric and all gas companies. While the latter was reported to have been a mistake, both were dropped according to Beck because of a storm of protest.[48] The second was that the municipalities lost the power they had held under the Ross Power Act to nominate commissioners. The third was that all contracts with the HEPC were subject to the formal approval of cabinet.[49] In sum, the provisions that the HEPC's growth was determined by demand in the municipalities and that its debts were repayable by the municipalities gave support to the notion that a municipal cooperative of sorts had been established, despite the extensive powers conferred on the government. However, the municipalities were given no control over the HEPC's decision-making even though they had initiated the movement for cheap power.

3

Ruling the Hydro-Electric Power Commission through Personal Leadership: The Beck Era, 1906–25

The Beck era of the Hydro-Electric Power Commission gained its renown from the mark Adam Beck left through his personal leadership. Serving continuously as chairman from the appointment of the first commissioners in 1906 to his death in 1925, Beck led and won two legendary battles that irreversibly altered and heightened the institutionalized ambivalence of the HEPC. First, he championed crown corporation autonomy for the HEPC at the expense of departmental controls, abetting the notion that the commission was the trustee of a municipal cooperative rather than a government enterprise to aid in the process. Secondly, he drove the private sector out of the electric power industry, for all intents and purposes transforming the original co-existence into a public ownership monopoly. By the time he died, his redistributive project had built the HEPC into a first-generation crown corporation broadly facilitative of economic growth.

Beck's strategy for securing corporate-like autonomy for the HEPC was most evident in his unwillingness to have it tacitly remain, let alone officially become, a department of government. Although he served for most of this era simultaneously as an MPP, cabinet minister, and commission chairman, this did not provide an enhancement of government control. When refusing to abide by the government's policy dictates, Beck balanced the government, his political and financial master, against his municipal allies, whom he readily mobilized against government control when necessary. He was able to exercise this independence because he had earned the undying allegiance of municipal leaders by promoting the myth that the HEPC was the trustee of a municipal cooperative. With department-like control diminished and the notion of a municipal cooperative popularized, Beck secured the autonomy of a crown corporation for the commission.

In fighting for the HEPC's autonomy, Beck did not simply seek to obviate government control. As a business it experienced such phenomenal growth in the twenty years following its inception that it required corporate-like financial practices. This growth resulted from Beck's unrelenting commitment to expand public ownership even though the commission had been created ostensibly to transmit power to municipalities in southwestern Ontario which the private sector would not serve from Niagara. Following an unlikely strategy, he led the HEPC and the municipalities into direct competition with the private sector on generation, transmission, and distribution. His objective was to undermine its monopoly-based worth in order to negotiate its subsequent purchase from a better position. Thus, he neither expropriated private companies nor nationalized them when their owners were ready to relinquish them, and he resisted other accommodations to this end originating within the government.

Public and Private Sector Co-existence

The appointments of the first commissioners were made on 7 June 1906, a month after the creation of the HEPC. The obvious candidate for the chairmanship was Beck, who had been a minister without portfolio since the Conservatives became the government in 1905. Indeed, it was likely that municipal leaders would have objected to anyone but Beck, given his leadership in the municipal movement and his role in two inquiry commissions which led to the HEPC's creation. To balance Beck's enthusiasm, Premier Whitney chose John Hendrie, a minister without portfolio like Beck, as the second commissioner, knowing he was associated with the private power interests in Hamilton. Hendrie resisted the call initially because he felt it would cause friction, but he accepted after Whitney assured him that the HEPC commissioners were to operate like a 'committee of cabinet.'[1] For the final appointment, Whitney selected Cecil Smith, who had been the engineer for Beck's inquiry commission. He had also been the resident engineer of the Canadian Niagara Power Company and the general manager of the government's Temiskaming and Northern Ontario Railway. Although the HEPC's membership left Beck subject to cabinet control approaching that of a government department, it led him to secure a power base outside of government.

The establishment of the HEPC had served to quiet longstanding public objections to private monopoly primarily because its intended mission was to transmit power to municipalities which the private sector did not plan to service. The Power Commission Act, however, contained no such restriction nor any rules on how the public and private sector were to co-exist.

Whether co-existence would involve segregated territory or direct competition mattered a great deal because the electric power industry had the characteristics of a natural monopoly.[2] The answer came within two months when the municipalities interested in contracting with the HEPC for local distribution, many of whom planned to compete with existing private companies, forged ahead without regard to their impact. On 24 July 1906, with Beck and Smith but not Hendrie in attendance, the municipalities formulated a standard bylaw for approving provisional contracts with the commission in the next municipal election.[3]

The municipal referendums on 1 January 1907, all of which succeeded, established the HEPC as a business in two important ways. First, they served as Beck's mandate to tender for a block of power from the private generating companies at Niagara and to contemplate constructing HEPC transmission lines to the contracting municipalities. In doing so, Beck set the vertically integrated Canadian-owned Electrical Development Company (EDC), the focus of past monopoly resentment, against the American-owned Ontario Power Company. The EDC had begun to transmit power to its Toronto distribution and street railway companies from its generating plant at Niagara the month before, only now to find that it faced competition. In building this plant, the EDC had justifiably presumed that the Ontario Power Company would continue to concern itself with exporting power to the United States rather than supplying the Canadian market. Secondly, the municipal referendums initiated unregulated competition between the public and private sectors. With this outcome most evident in Toronto, a market which both sectors needed for viability, the EDC was left in an even more precarious position. The fate of the company, however, was not yet settled. The second stage of the municipal approval process, authorizing debentures for local distribution works, would not be reached until 1 January 1908.

Meanwhile, during the winter of 1907, two alterations were made to the HEPC as constituted in 1906. After seeing the commission through its first year and having assisted Beck in assembling a permanent engineering staff, Smith resigned on 28 February to become a consulting engineer. He was replaced by W.K. McNaught, who had been a strong advocate of public power while a member of the Toronto Board of Trade and had since become a Conservative backbencher in a 1906 by-election. Beck also had the whole Power Commission Act passed again on 20 April, but this time strengthened the HEPC's capacity to construct its own facilities and purchase power generated by private companies. The new act also left no doubt that municipalities contracting with the commission were exempt from the Conmee Clause of the Municipal Act. For James Mavor this enactment and others which followed were part of a deliberate policy to frighten away capital from the EDC so as to impair its viability.[4]

When the result of the HEPC's tendering for a block of power became known, it raised the first of many public/private sector controversies. The fact of the matter was that the Ontario Power Company had agreed to sell power cheaper than the EDC. To stave off the ruinous financial consequences of this outcome for the EDC, Whitney approached Ontario Power privately with a proposal for a shared contract. The company rebuffed this overture, believing that the EDC's principals sought to 'make themselves millionaires no matter what the cost might be to others.'[5] Although Beck was permitted to sign a contract with Ontario Power on 30 April 1907, Whitney's manoeuvring nonetheless perturbed him; he had already publicly committed the commission and the municipalities to open competition on transmission and distribution with the EDC. The experience provided the first evidence that commission operations would not exclude political considerations.

The HEPC's affairs were destined to remain enmeshed in politics because its existence, let alone any future success, did not bode well for the financial viability of the EDC. To forestall the EDC's demise, Whitney secretly continued to negotiate with the EDC and Ontario Power, this time for formal government regulation of territories for transmission. He did so knowing that this would require overriding some municipal votes, including Toronto's, and Ontario Power's signed contract with the HEPC. On 4 October these negotiations also failed, but this time because the EDC refused to submit to government regulation. The company believed that regulation was a breach of its 1903 franchise agreement with the Queen Victoria Niagara Falls Park Commission. The contract had a provision stating that the government would not generate power, leading the EDC to claim that regulation promoted such an alternate source of generation.[6]

Despite the EDC's intransigence with Whitney, the exemption from the Conmee Clause led it to see regulation as protecting, rather than threatening, its interests. The reason was that this clause ensured that the second stage of the municipal bylaw referendum process – approval of debentures for municipal distribution works – provided a proper legal sanction for unregulated public/private competition. To avert the prospect of local competition, the EDC principals proposed three alternatives: two levels of regulation, and the sale of its Toronto Electric Light Company to the city. With all three conditional on the city's being bound to the company's long-term contract with the EDC and the sale including the value of its franchise which ran to 1919, the conditions scuttled any chance of a deal. The EDC's fate then fell to the electors of Toronto on 1 January 1908. They authorized the city to construct distribution works with connections to HEPC transmission lines in direct competition with Toronto Electric Light as supplied and controlled by the EDC.

Private Ownership and Government Regulation

The failure of the EDC's principals to avert competition on both transmission and distribution led representatives of its bondholders to approach Whitney directly in January 1908 with new pleas for private ownership *with* government regulation. This outcome was essential to straightening out the EDC's financial problems, but politically the appeal, coming after the second municipal bylaw referendums, was too late. With regulation now ruled out, the EDC's joint-stock ownership was brought to an end. Its total collapse was only averted when one of its principals, William Mackenzie, bailed out the company on 14 February. He created a holding company, the Toronto Power Company, to administer all of the vertically integrated properties, after refinancing the EDC through guarantees on his street railway properties.

Despite Beck's initial victories in promoting the HEPC, Mackenzie believed that increased government regulation would prevail over public/private competition. He based this on a discussion he had had with Whitney in the summer of 1907. In the winter of 1908 he wrote to the premier proposing that the Toronto Power Company and the Ontario Power Company jointly own existing and new transmission lines and make contracts directly with municipalities. The government, for its part, would regulate the location of lines and the rates charged and guarantee the bonds of private companies in order to keep costs down. Whitney, however, never officially responded to Mackenzie's proposition. He likely recognized that any decision favouring regulation was complicated by the fact that it would also serve as the government's decision not to permit the HEPC to build its own transmission lines. Given that he was also contemplating an election, it was not surprising that no immediate decision was made on this politically explosive issue.

Knowing that the plight of the defunct EDC had previously garnered sympathy from the government, Beck acted to strengthen the HEPC's hand with the government for the new round with Mackenzie's Toronto Power Company. He did so through the commission's official thirty-year contracts with the municipalities, dated 8 May 1908. Besides their preamble speaking of the HEPC as 'acting on its own behalf with the approval of the Lieutenant Governor in Council,' they promoted the view that the HEPC was a municipal cooperative rather than a government enterprise. Specifically, they stated that 'the [Hydro-Electric Power] Commission is to be a *trustee* of all the property held by the Commission under this agreement for the [municipal] Corporations and other municipal corporations supplied by the Commission' [emphasis added]. In addition, the billing scheme outlined in the contracts further buttressed the cooperative contention. It provided for twelve monthly payments at a fixed rate with an annual or thirteenth bill for adjustments and apportionments, an arrangement that would later be heralded as 'cooperative accounting.'[7] Given that these cooperative provisions would later be included in all other municipal contracts, Beck had laid the foundation for a powerful myth that the HEPC was not a government enterprise. More immediately, the municipal backing the cooperative contention received made it difficult for the government to accept Mackenzie's proposal for regulation.

Whitney was not in a position to impose regulation of private enterprise in place of public/private competition because he was not well situated to discipline Beck to this outcome. The reason was simply that Whitney wanted Beck's support in the June 1908 election, a public endorsement which he received. Following the election, Whitney decided against regulation, but his decision only became evident when contracts were signed on 13 August for the construction of HEPC transmission lines. This left Mackenzie feeling betrayed. He claimed that he had taken financial responsibility for the EDC only on the basis of Whitney's private assurance that there would not be public/private competition on transmission of power. Although Whitney had not publicly rejected Mackenzie's offer, he denied giving the assurance. With the Toronto Power Company's animosity now directed at himself, Whitney recognized there was no turning back on HEPC transmission lines without 'practical ruin and humiliation' for the government.[8]

The decision to permit the HEPC to construct transmission lines was followed in the fall of 1908 by legal challenges to the validity of the contracts the cities of Toronto and London had signed with the commission. At issue was a discrepancy between the bylaws passed by voters in 1907 and 1908 and the contracts signed with the HEPC. These actions were mounted by local petitioners associated with Toronto Power even though the government, with the election looming, had remedied this discrepancy by retroactively validating all the municipal bylaws and the commission's contract with Ontario Power by statute in May 1908. By refusing to grant leave for the HEPC to be included in these legal actions, Whitney, acting in the absence of J.J. Foy, the attorney general, hoped the controversy would be at an end. However, the government had to respond by statute again when the mayor of Galt still refused to sign a contract with the commission. In 1909 it legislated the Galt contract as 'executed' and placed all HEPC contracts beyond the jurisdiction of the courts.[9]

Toronto Power reacted to this prohibition on legal challenges by petitioning for federal disallowance. Although the petition claimed that there was a federal interest in the matter because the Niagara River was an international boundary, the company hurt its case by also stating that the Ontario legislation was unconstitutional in a political rather than a legal sense. The Ontario government countered that there was no federal interest in the matter because the river was completely unnavigable and stated that power developments on the river could only fall under provincial jurisdiction of property and civil rights. In addition, Beck responded to the challenge. Through amendments to the Power Commission Act on 19 March 1910 he secured for the HEPC the power to regulate the construction and operation of municipal as well as private distribution and transmission works.[10] On 29 March Prime Minister Wilfrid Laurier announced that he would not invoke disallowance even though he found the 1909 legislation objectionable. Since the Ontario action was unjust rather than *ultra vires*, his hands were tied.[11]

In the immediate aftermath of this decision, Whitney counselled the parties to resolve their differences. The Toronto Power Company then offered to sell the Toronto Electric Light Company to the City of Toronto, but no deal could be reached. Again, the price was based on its value without competition and the deal included its long-term contracts with the other companies.

The competition between the public and private sectors was consolidated with the completion of the HEPC's transmission lines to contracting municipalities, beginning with Berlin on 11 October 1910. Despite this milestone, Whitney acted to terminate the commission's 1910 contract with the City of Hamilton and remove its new power to regulate municipal and private distribution and transmission works. His reason, undoubtedly, was that Beck's contributions to the protracted public/private sector conflict reflected negatively on his ability to manage the government. Through a bill introduced on 25 January 1911, Whitney wanted to place the regulatory power in question with the Ontario Railway and Municipal Board. Besides usurping his authority, Beck was sceptical of these two changes for another reason. Hendrie, whose function on the commission was to keep Beck in check, was associated with the private power interests in Hamilton, had led the creation of the Railway and Municipal Board in 1906, and was chairman of the legislature's railway committee. In the resulting clash, Whitney did not proceed with the bill, allowing it to be withdrawn on 20 February at second reading.[12]

Whitney also raised the issue of the HEPC's relationship with the government when he called the December 1911 election. He declared that the HEPC should be 'discontinued' and replaced with a 'Department of Power,' in this way seeking to capitalize on its popularity throughout the province. This raised the ire of the existing municipal hydro commissions, especially because their champion to date, Beck, was strangely silent. Beck was in fact against Whitney's pronouncement, Nelles reports, because it would place the HEPC's 'policy and expenditures under the scrutiny of the full cabinet and, therefore, into the range of all other government programs and spending priorities.' However, Beck did not rebuke his leader, according to Plewman, because he was unsure of his own strength and did not want to cause his departure from his party and government. The election, which gave Whitney a third consecutive large majority, was widely recognized as an endorsement of province-wide government power transmission.[13] While the best form for its administration was not equally clear in the public's mind, events prompted Whitney to take no action on altering the HEPC's status.

Although he did not challenge Whitney's pronouncement either during or after the election, Beck benefited from the parliamentary and extra-parliamentary opposition which developed to the plan. The new Liberal leader, Newton Rowell, had come to the defence of the HEPC's existing structure both to keep patronage out of its affairs and to protect the rights of the municipalities, whom he considered to be its real owners. Whitney's declaration also led municipal leaders to protest that the plan constituted confiscation of their cooperative ownership of the HEPC. Their case was especially strong given that twenty-nine new municipalities voted to contract with the commission in 1 January 1912 referendums. Given this opposition behind him, Beck was able to announce on 11 January that the HEPC would not become a department of government. However, as a face-saving device for Whitney, Beck's proclamation was qualified to stand only while the HEPC was localized to the Niagara region. Merrill Denison writes that with Whitney's retreat, the commission became quasi-autonomous.[14]

In the month following Beck's announcement, the HEPC's new autonomy gained formal interest group support. In February 1912 the municipal movement left dormant following the creation of the HEPC in 1906 was reconstituted as the Ontario Municipal Electric Association (OMEA) with the explicit objective of supporting Beck. Although there had been rapid growth in the number of municipalities which had contracted for power since 1907, their collective voice had not been kept organized. The OMEA addressed this challenge in its statement of purposes. Advocating the view that the HEPC was the trustee of a municipal cooperative, it called for united action in promoting the 'interests of the municipal electric undertaking of the Province,' particularly by working in conjunction with the commission and by proposing legislation of advantage to the undertaking as defined. As a result, the OMEA provided Beck with what Nelles describes as bipartisan, extra-parliamentary authority. It would also give Beck, in Plewman's view, unfailing support both as a sounding board for his policies and as a political weapon in his battle to keep the HEPC free from the government's dictates. The OMEA's confidence in Beck was unbounded.[15]

One of the OMEA's first acts was a call on the government to provide Beck a salary commensurate with the responsibilities of a chairman of a large utility. The Power Commission Act was amended accordingly on 16 April 1912 to permit the cabinet to set a salary of up to $6,000, notwithstanding the prohibitions on such payments in the Legislative Assembly Act. This salary was in addition to the $6,000 remuneration Beck received as a cabinet minister and $1,400 he received as a sessional indemnity. While the salary provision was an important move toward a corporate-like organization, it did not alter the form of the HEPC's department-like subservience to government. Beck's new salary was to come out of appropriations, as did all commission expenditures, rather than revenues, which were still to be turned over to government in full. The act was also amended to strengthen the commission's powers to regulate transmission and distribution matters, this time explicitly in lieu of the Ontario Railway and Municipal Board, and to reconfirm Hamilton as a HEPC municipality. During debate on this legislation, Rowell added a novel twist to the increasingly confusing issue of the commission's ownership. He argued that the municipalities should have representation on the HEPC.[16]

The HEPC continued to receive criticism for its very existence, despite a reprieve from the Toronto Power Company. A joint committee of the New York State legislature chaired by Senator Harvey Ferris commented in its first report on water power in January 1912 that regulation would have served the same purpose as public ownership in Ontario. Then, in a second report in January 1913, undertaken as a cost-benefit analysis of adopting the HEPC scheme in New York State, it concluded that the commission was a failure by 'business and economic standards.'[17] This report attributed the failure to supplying power below cost and thereby belittled the HEPC benchmark of at-cost power. Although the report surmised that the failure was a result of pressures to prove that the system could deliver cheap power, and thus keep favour with the public, the effect of its criticism in Ontario was not entirely negative. It drew attention to Beck's essentially populist operation of the HEPC, the factor that made it such a political success.

The Ferris Committee's criticism was expanded upon and given wider dissemination in a 1913 book by Reginald Pelham Bolton, a New York engineer who had greatly influenced the committee's 1913 findings. His general point, like that of the first Ferris report, was that the public interest would have been better served by regulation of rates and that public credit should not be used to force competition where it was 'naturally unsuited.' His more damning criticism was that the HEPC's rates were 'arbitrarily and adroitly' set to undermine private companies. Although he claimed to represent no special interest, Bolton, in Plewman's view, had some powerful business backers, presumably fearful of the spread of public ownership. What Bolton saw as unprofitably low rates, however, the HEPC regarded as a promotional rate structure. In essence, the commission maintained that it used low rates to stimulate 'greater economy of operation' through increased load density and then passed on the savings to consumers to continue the cycle.[18] Although Bolton may have regarded this as uneconomic consumption, the low-rate strategy won immense public approval.

The success of the HEPC's rate strategy increased Beck's capacity to shape the character of the electric power industry in Ontario. From this position he chose to argue against a rationalization of the industry through nationalization. This was made evident in 1913 when Whitney considered purchasing the Toronto Power Company's generation and transmission companies and the City of Toronto sought to purchase the company's distribution and street railway companies. Beck objected to both schemes because the purchases would be made from weak negotiating positions, causing prices to be high. He wanted to wait until the Toronto franchise agreements expired in order to remove the market for Toronto Power's generation and transmission companies. The purpose was to undermine the value of the whole package. Thus, Beck felt that competition would, in the end, rationalize the industry. According to Mavor, this was an insidious policy because it was sustained by the HEPC's capacity to set rates at a point where their consequences would be ruinous for private companies.[19] Whether the rate was arbitrary or promotional, Beck was gaining the influence he coveted to shape the industry.

Although the political popularity of the HEPC had been good for the Conservatives, the government was reluctant to support Beck's ambitious plans for the commission. Beck wanted a stronger government commitment to inter-city electric railways than he had received in the Hydro-Electric Railway Act of 1913. While he promoted these 'radials' publicly to address the demands of urbanization, they also served his interests by increasing the load on the HEPC's transmission system. He also wanted to move the commission into the field of generation, particularly through competition with private companies at Niagara, leaving behind its self-imposed (rather than statutory) limitation to transmission. Finally, Beck insisted on a division of labour in the public sector where the provincial commission controlled transmission and generation, restricting the municipal commissions to local distribution of HEPC power.

This last demand surfaced as a result of the decision of Toronto Hydro, which was still in competition with Toronto Electric Light, to reject rate reductions ordered by the HEPC. Toronto Hydro was intent on raising capital ostensibly for a back-up steam generation plant of its own. Although its 1908 contract with the commission permitted it to generate power in emergencies or to reduce peak demand, Beck felt that the plan trod on the HEPC's responsibility. In this dispute, it was notable that one of the three Toronto commissioners, R.G. Black, was formerly general manager of Toronto Electric Light and the chairman, P.W. Ellis, had opposed Beck over provincial government involvement in the electric power industry in 1906. Given these machinations, Beck also stood adamantly against to the plan because he felt it would benefit the Toronto Electric Light and hurt the integrity of the HEPC.[20] When the government hesitated to do his bidding on this and his other two demands, he threatened to bring it down, forever leaving behind his weak stance of 1911–12.

The government heeded one of Beck's wishes by passing a new radials act in the spring of 1914. It also amended the Power Commission Act on the recommendation of the OMEA to provide for combined salaries of up to $15,000 for the commissioners. Rather than take political responsibility for paying him out of appropriations, however, the government established that the new salary compensation be raised by municipal levy. It thereby bolstered the notion that the HEPC was a municipal cooperative. The funds were divided up in such a way that the chairman received $6,000 and the two commissioners $4,500 each, meaning Beck now received $19,400 in total for his commission and government services.[21] He nevertheless remained dissatisfied. But with an election looming, the government did not risk alienating him as it had done with the proposal for a department of power. During the campaign for the June 1914 election, Whitney promised to support Beck's other two HEPC demands and ensure that he received a knighthood. This paved the way for Beck to endorse the government, which went on to win a fourth consecutive large majority. In the process, Beck had weakened the government's department-like political control over commission policy and, along with the salary provision, had strengthened its growing corporate autonomy from the government.

As for Whitney's two promises, the government did permit the HEPC to enter the field of generation in 1914, although not through competition or nationalization at Niagara. Having been successful as a transmission company since 1910, the commission completed the construction of its first small generating station in 1914 and purchased other existing small plants shortly after. With these plants located north of Toronto and needed primarily to meet HEPC commitments to new municipal hydro commissions out of reach of Niagara, they did not satisfy the growing demand in the existing serviced municipalities where war mobilization now exacerbated the problem. For the latter, Beck would have to expand power purchases from the existing private companies at Niagara in 1915 while waiting for Whitney's commitment to permit the HEPC to generate power there to be honoured. The second promise, regarding the provincial-municipal division of labour in the public sector, would not be acted on until the 1915 legislative session.

After the 1914 election three departures from cabinet further eroded the commission's department-like relationship with the government. Whitney, who was seventy-two and in failing health, died on 25 September. Then, following Whitney's death, Beck resigned from cabinet, claiming the commission demanded his full-time attention. Lastly, Hendrie, the other cabinet minister who had been a commissioner from the HEPC's inception, resigned that September on being named lieutenant-governor of Ontario. The only other commissioner, W.K. McNaught, was not in a position to renew the departmental connection to the government. Not only was he not in the cabinet, but at age sixty-nine he had not stood for re-election to the legislature. In sum, a vacuum had been created in the political control capable of being exercised over the HEPC, and it was likely that it would never be filled to the same degree while Beck remained chairman.

With Whitney and Hendrie no longer around to restrain him, Beck's desire to operate the commission with corporate-like business autonomy depended only on the disposition of the new premier and the new appointee to the HEPC to replace Hendrie. William Hearst, the member for Sault Ste Marie who succeeded Whitney as premier on 14 October, was not well positioned to challenge Beck. Hearst had had little opportunity or reason to familiarize himself with the HEPC. Having been first elected in 1908, he had not been party to its creation, and having come from the north and been minister of lands and forests from 1911 to 1914, he had no direct connection to the commission. To make matters worse, Beck had been passed over for the premiership in favour of Hearst. The new premier's weak standing with Beck meant that the attitude and stature of the new appointee to the HEPC, who by statute had to come from cabinet because Beck and McNaught did not, would have a critical influence on whether a strong department-like linkage would continue to exist with the government. Although I.B. Lucas, the provincial treasurer who would become attorney general in December, was appointed, his ability to make this connection was complicated by the fact that Beck remained in the legislature. According to Nelles, Beck began to suggest at this time that the HEPC was directly responsible to the legislature through him as chairman.[22]

Beck succeeded initially in getting his way with the Hearst government. His dispute with Toronto Hydro over rate reductions and municipal generation was resolved through amendments to the Power Commission Act during the 1915 legislative session. Although the amendments did not explicitly dictate that municipal hydro commissions could not generate their own power, they accomplished the same objective by less heavy-handed means. The act now made it explicit that the municipalities and their hydro commissions were duty bound to uphold the terms of their long-term contracts with the HEPC, with the latter able to seek court injunctions to enforce the provisions. The significance of this amendment was that the contracts signed in 1908 and afterwards stated that municipal hydro commissions had to take power 'exclusively' from the HEPC, the only exceptions being emergency situations and meeting peak load requirements. The act now also permitted the council of a municipality with a population of over 100,000 to create an appointed (as opposed to elected) commission of three consisting of the mayor, one council appointee, and one HEPC appointee. When Toronto Hydro was subsequently organized as such, the HEPC was able to exercise informal influence in its affairs. Finally, undoubtedly triggered by Hendrie's associations in Hamilton and the presence of Black, formerly of Toronto Electric Light, sitting as a Toronto commissioner, strict provisions were established to bar persons with financial interests in private electric companies and manufacturing industries from holding appointed or elected positions with local hydro commissions.[23]

Forging Corporate Autonomy

In response to Beck's domineering manner and his operation of the HEPC as if it had corporate autonomy, the government set out to constrain him on three fronts in 1916. According to Peter Oliver, the actions were taken in public because Hearst was at a loss as to how to deal with Beck privately. For starters, Attorney General Lucas secured amendments to the radials act that halted bond issues and construction on Beck's cherished scheme until after the war. Although this particular constraint occurred with the least fanfare, Beck nonetheless believed that radials would not take capital and materials away from the war effort. On the two other fronts, Oliver reports, Hearst unleashed his cabinet 'firebrands,' Howard Ferguson, the minister of lands and forests, and Thomas McGarry, the new treasurer.[24] Ferguson undercut Beck's general strategy of competing with private companies while waiting out their leases, and McGarry attempted to end Beck's management of the HEPC's finances, contrary to the Power Commission Act, in a corporate rather than departmental manner.

As a politician from eastern Ontario, Ferguson had been anxious to advance the expansion of the HEPC into that region as early as 1912, but Beck did not appreciate his interest in 1916. Publicly, Beck held that the municipalities in question had not expressed their displeasure with the existing private companies, as had others, by voting to contract with the commission. A more plausible reason for his position, of which Ferguson must have been aware, stemmed from the fact that Beck had unsuccessfully attempted to purchase the Central Ontario Power Company, owner of the numerous local systems, in 1915. When the purchase offer was refused, Beck decided to enter into direct competition with the company. At this point Ferguson established a different agenda, although one complicated by intransigence on Beck's part. Without the participation of the affected municipalities, Ferguson had legislation passed that April authorizing the Ontario government to purchase the company and the HEPC to operate it on the government's behalf. Beck accepted the company on these terms only reluctantly, believing that the local municipalities would not accept the debt obligations that stemmed from the high price Ferguson paid.[25] Indeed, the system would not be incorporated into the commission until after Beck's death.

McGarry, for his part of the assault on Beck, set out to reverse the loss of cabinet control over the HEPC's finances. This was a problem which had been building for many years, but one which had been exacerbated by

Beck's departure from cabinet in 1914. Although Beck did mitigate the problem somewhat by continuing to explain the HEPC's estimates in the legislature, the overall arrangement was less than satisfactory because he exerted little effort to keep the cabinet informed. The problem was brought to a head for McGarry when James Clancy, the provincial auditor, appended an unusual but forthright note to the figures he included for the commission in the 1914–15 public accounts, submitted in January 1916. It stated that the balance sheet in question was 'prepared by the HEPC and does not represent the accounts of the Commission as audited by the Audit Office.'[26] Here began a protracted dispute over the commission's finances. At issue was the propriety of their administration in a corporate fashion.

Clancy's note was followed by a special report he prepared for McGarry on 21 February. Here Clancy maintained that an audit of the HEPC for the entire period of 1909–15 had been impossible to complete because the commission had failed to furnish complete accounts. To make matters worse, he claimed that the HEPC, in defiance of the Power Commission Act, had charged the government for unauthorized expenditures; substituted itself for the municipalities as the sinking fund debtor; undertaken commercial projects outside its responsibilities; and set power rates above cost to generate and retain surplus revenues. The last was a very serious charge, involving expenditures of $4 million more than the $13 million the commission had received in appropriations. It indicated that Beck had been applying the HEPC's revenue directly to expenditures when he was required by statute to turn all revenues over to the treasurer and make expenditures out of appropriations. Clancy attributed the problems to two causes: 'One, the absence of even a semblance of legislative control over the expenditures of the Commission – in striking contrast with the complete legislative control over the Executive Departments. The other, the seemingly defiant disobedience of the act creating the Commission with their powers and duties.'[27]

Unwilling to stand rebuked for his administration of the HEPC's accounts, advising accountant W.S. Andrews joined the fray through a report to Beck on 4 March 1916. In contrast to Clancy's strict reading of the Power Commission Act, he remarked that the commission's accounting practices were based on a broader interpretation, one read against the municipal contracts as also approved by the legislature. He added that the government had been in full knowledge of these practices and 'advised of the deficiencies of the Act.' On the matter of disclosure, Andrews believed that the provincial auditor's role was only to check the veracity of the HEPC's books, whereas Clancy wanted to see every voucher to perform a full audit. Moreover, Clancy's specific charge was a surprise to Andrews because no such audit had been attempted until late 1914 and then again in late 1915, whereupon the commission fell under intense scrutiny and pressure. Indeed, on the latter occasion, McGarry extracted the vouchers desired by Clancy by refusing to forward requisitioned funds. As a retaliatory rebuke to Clancy, Andrews closed by stating that the report to McGarry contained 'several inaccuracies in the additions pointing to careless preparation.'[28]

The dispute between McGarry and Clancy on the one side, and Beck and Andrews on the other, dominated the 1916 proceedings of the legislature's public accounts committee. In testimony on 29 March, Clancy admitted that he had no need to examine the commission's books because 'executive departments,' in which he included the HEPC, do not have their own assets and liabilities. Beck retorted that Clancy's position was 'an impossible condition from a business standpoint.' In later testimony Andrews similarly belittled Clancy's view, arguing that he had 'never learned to keep accounts without books' and that Clancy sought to 'duplicate our books without keeping books.'

In the absence of any agreement on why the audits had not taken place, Clancy, who had been a Conservative MPP from 1883 to 1894, raised the dispute to a constitutional issue, stating: 'So long as we live under the rule of law the Legislature must be the supreme power in government and the Commission has assumed the power of the Legislature in attempting to distribute [burdens upon individuals or upon municipalities or upon this province].' Then, demonstrating the unworkability of his view for a business the size to which the HEPC had reached, he added: 'It is impossible to have legislative control unless the Legislature takes into its own hands the spending of monies for the precise purpose it intends.' After this lecture to the assembled politicians, McGarry's enthusiasm for defending Clancy in the proceedings waned. He brought the controversy over disclosure and the administration of accounts to a end for the time being by announcing to the committee on 11 April that he had hired noted chartered accountant E.R.C. Clarkson to conduct an independent audit. Beck heartily welcomed the announcement.[29]

Not surprisingly, Beck had remained unrepentant in the face of the provincial auditor's findings. He defended his operation of the HEPC on the grounds that the control outlined in the Power Commission Act, including the department-style audit, was unworkable. Thus, he not only accepted personal responsibility but also countered that the commission required corporate-like business autonomy. To facilitate this outcome, the OMEA put pressure on the government for direct representation on the HEPC, likely on Beck's instruction. Given that this would have reduced the government's and the legislature's already minimal control, the proposal was rejected. Beck's desire for autonomy was nonetheless later supported in Clarkson's 1918 report to McGarry. It recommended that the Power Commission Act be amended to conform with the existing financial practices, rather than the reverse, as McGarry had originally sought.[30]

In the end, the amendments McGarry made to the Power Commission Act in April 1916 reflected a compromise with Beck. Of the two groups of amendments regarding the administration of the HEPC's accounts, one set allowed corporate-like financial freedom and the other augmented the commission's financial accountability to the government.[31] On the first, the HEPC was permitted, following the Clarkson report's outline, to establish a reserve fund for renewal of its works. This meant that the commission could now apply its revenues, except those from the municipal hydro commissions specifically devoted to sinking fund and interest, directly to expenditures. This was a significant concession because it permitted corporate-like financial practices. It was partially offset, however, by the provision that appropriations were now to be paid over only after specific requisitions received cabinet approval. Consistent with this autonomy, the HEPC could now select its own auditor, among whom the provincial auditor was eligible, for an annual audit. Taking a different view, the Liberal opposition had argued that the hydro municipalities should appoint the auditor.[32]

The second set of amendments, those to increase government supervision, centred on a new requirement for the HEPC to submit an annual financial report. Although the commission had done so informally since 1908, the Power Commission Act now specified that the annual report must disclose both current and expected assets and liabilities, income and expenses, receipts and disbursements, and securities and indebtedness. By containing these features, the report appeared to be a marker for corporate autonomy, but on balance it was an instrument of department-like control. The reason was that the cabinet was given the power to appoint a HEPC comptroller whose statutory responsibility was to supervise the commission's financial records and submit the annual report to the cabinet for the 'information' of the legislature.[33]

Not surprisingly, Beck and the OMEA rallied against the appointment of a comptroller. Their opposition led to another compromise being struck in a 1917 amendment, in effect undoing the 1916 amendment. This one lessened the comptroller's supervisory power to that of making recommendations to the HEPC commissioners and permitted the commission to nominate the comptroller with the cabinet left only to approve the appointment. Even with these concessions, however, the requirement that the annual report be submitted through the comptroller was repealed in 1918. And, without a comptroller ever having been hired, the function was later dropped when the Power Commission Act was consolidated in 1927. Moreover, the 1918 amendments to the act strengthened the contention that the HEPC was the trustee of a municipal cooperative. The act now required the annual report to detail the sinking fund contributions of each municipality, drawing attention to the fact that the municipalities, not just the government, had a financial stake in the commission. Beginning in 1919, these contributions were itemized, much to the government's chagrin in later years, as 'equity in the HEPC.'[34]

Beck had won the corporate autonomy struggles with the Hearst government in part because of the need for new sources of power to meet war demand. Knowing that only Beck was capable of delivering power quickly, the government opened new avenues for him to facilitate power production. Through statutory enactments in 1916 and 1917 the HEPC was permitted to pursue three courses of action: to enter the field of generation at Niagara, having had this decision held in abeyance since the 1914 election; to buy out existing plants; and to regulate the private sector.

The first avenue followed from the passage of the Ontario Niagara Development Act in 1916. This specified the terms on which the commission could proceed with preliminary work on its own generating station, one that would operate in direct competition with the private sector.[35] Hearst stipulated, however, that extraordinary referendums were to be held in the HEPC's contracting municipalities in the municipal elections on 1 January 1917 before further legislative approval would be forthcoming. Beck and the OMEA responded to the challenge by organizing campaigns that garnered the necessary public approval. In this regard, Hearst's demand bolstered the notion that the HEPC was the trustee of a municipal cooperative, but it also solidified for the government the public support necessary to counter private sector opposition. On 2 January Hearst promised the supplementary legislation for the planned Queenston-Chippawa power development, and Beck announced that the project would cost $30 million for the station and $20 million for the water diversion canal. At 500,000 horsepower, it would be the largest electric station in the world.[36]

The Ontario Niagara Development Act of 1917, supplementing the 1916 act, not only sanctioned Beck's scheme to compete with private companies on generation at Niagara but also increased the HEPC's autonomy from government. The commission was now permitted to issue its own bonds for the construction of the project and, subject to cabinet approval, to offer a government guarantee. This gave the HEPC an important marker of corporate autonomy by exempting the project from the normal departmental controls over raising capital. The act also specified that the commission, for the purposes of the works constructed under the act, was a 'trustee' for its contracting municipalities, although the government would hold a lien on the works. Both the bond and trustee provisions fitted Beck's persistent campaign to undermine government control, and his obstinacy on this matter knew no bounds. In 1917 he actually rejected a government claim that the HEPC reported to the legislature through Lucas, the attorney general, who was also a commissioner. This would have been only a small concession because the claim was accompanied by an admission that the province owned and operated the commission 'in trust' for the municipalities.[37]

While the mention of the word trustee enhanced the municipal cooperative notion, it was included in part to help insulate the government from private sector complaints. The Queenston project represented an assault on what had been the last preserve of the private sector, generating power at Niagara, and the passage of the supporting acts did not go without incident. The Toronto Power Company responded by launching a court action, basing its case on the provision in its franchise agreement that the government would not generate power. It did so even though the HEPC had received a statutory exemption.[38] When Lucas, despite the conflict of interest of his two responsibilities, refused to grant leave for the suit to proceed, Toronto Power petitioned for federal disallowance, arguing that the Ontario Niagara Development Acts of 1916 and 1917 were 'illegal competition.' Lucas reminded the federal justice minister, C.J. Doherty, of the 1909-10 precedent on the matter set by Laurier, and on 4 May 1917 Doherty refused to intervene.[39] Nevertheless, this would not be the company's last legal challenge.

The second avenue opened for the HEPC to attain new power sources, the buy-out of existing companies, came through amendments to the Power Commission Act in 1917. The HEPC was now allowed to acquire private companies through share purchases; to issue securities to pay for the shares; and to clothe the securities with a government guarantee if it had received cabinet approval. The ability to purchase private generation companies was critical to the feasibility of the Queenston project because Beck needed control of the Ontario Power Company's unused water rights for the planned water diversion. Thus, he acquired all the shares of the company in August for $22 million in HEPC securities.[40]

The third avenue opened for the HEPC to expand power production, government regulation of private generating capacity, led to another conflict with Toronto Power. The Water Powers Regulation Act of 1916, initiated by Beck, had permitted the cabinet to appoint an inspector to investigate all power producers for, among other things, the quantity of water available, permitted, and used, and the quantity of power available, permitted, and developed. The purpose was to enable the inspector to make orders for maximizing efficiency. In a conflict of interest not unlike Lucas's, Beck, who was appointed the inspector, had used his new role to extract 50,000 horsepower from the Canadian Niagara Power Company and charge the Toronto Power with producing 11,000 more horsepower than its franchise permitted. The latter answered in February 1917, objecting to the anomalous situation of being regulated by its competition.[41]

In response to Toronto Power's unwillingness to be regulated, Beck countered by securing a 1917 amendment that strengthened the Water Powers Regulation Act. This enactment permitted the cabinet to appoint a commission of three Supreme Court of Ontario justices to inquire into an inspector's charges. If they reported that a company had made an infraction under the act or had unused capacity, the cabinet could order the company to turn over power to the HEPC at a price determined by the judges.[42] This amendment was well crafted. Beck knew that Toronto Power had installed capacity of 25,000 horsepower more than its franchise permitted. However, when he tried to obtain this power, the company petitioned again for federal disallowance.

The company's petition was notable because it complained, among other things, that the legislation in question, and historically all HEPC legislation, had been purposely held back until the end of the session to prevent debate. The complaint was well founded, Plewman suggests, because the practice 'would not have happened if Beck had desired it otherwise.' Lucas's formal response to the petition argued not only that the acts were *intra vires*, but also that the company had illegally produced more power than permitted. Doherty, the federal justice minister, immediately refused to disallow the 1916 act (which had provided for the inspector) and, after reserving judgment, reported on 24 April 1918 that he could not recommend disallowance of the 1917 amendment. The next day the Ontario cabinet ordered Toronto Power to deliver the power to the HEPC. The cabinet made the decision, Nelles reports, on the recommendation of an inquiry hand-picked by Beck.[43] It included Chief Justice William Meredith, Whitney's predecessor as leader of the Ontario Conservative party and the ghost author of the Power Commission Act.

Despite the intractability of their positions, the years of bitter competition between the HEPC and the Toronto Power Company neared an end in 1918. In that year, negotiations began for the former to buy the latter's properties, although neither side appeared too anxious. The company wanted $45 million, but Beck offered only $27 million.[44] He did so on the advice of chartered accountant Geoffrey Clarkson, now in charge of his father's firm and an economic sage for the commission and the government. Although no deal was worked out, it was clear that one would be forthcoming. The company needed to sell before its Toronto distribution and street railway franchises ran out in 1919 and 1921, respectively. If it did not, their value and that of the company's generation and transmission companies would be severely diminished. For its part, the HEPC needed to buy the Toronto Power Company because it was facing a power shortage that would not be met before the Queenston-Chippawa development could be completed.

Reining in Beck's Personal Leadership

Although the war had expedited Beck's plans for generation at Niagara, it had caused his plans for radials to be put on hold. They were only revived when the 1916 restrictions on financing and constructing radials were repealed in 1919 following the war.[45] Despite the decline in earlier public enthusiasm for radials, Beck forged ahead, openly blaming the changed prospects on a lack of commitment by Hearst. Undoubtedly aware that a conflict with Beck would not bode well for his government, then stretching into its fifth year, Hearst responded that he had only been respecting Beck's dictate that the HEPC be kept out of politics. The conflict nevertheless continued and contributed to the government's defeat.[46] In the campaign for the October 1919 election, Beck had sent an unmistakable signal to voters on the radial issue when he bolted party ranks to run as an independent hydro candidate. As a result, the United Farmers of Ontario (UFO), a new party, won forty-five seats and formed a coalition government with eleven Independent Labour Party (ILP) members, leaving twenty-nine Liberals and twenty-five Conservatives in opposition. Beck, Hearst, and Lucas were all personally defeated.

Beck's unrelenting desire to operate the HEPC with corporate-like autonomy eventually led the UFO government to attempt to undermine him, but only after it had failed to bring him onside. Searching for a premier after having won the election without a leader, the UFO gave Beck consideration, but the parties differed over the autonomy of the HEPC. Thereafter, the premier chosen, E.C. Drury, asked Beck to join his cabinet with portfolio for power, but Beck rejected the offer, as he had in the past, as a matter of principle.[47] Although this rejection suggested that Beck felt he was unassailable, he was concerned about his tenure as chairman. Indeed, he orchestrated at this time an OMEA call to fix the chairman's term at six years (one year beyond the maximum possible life of the UFO government) and to expand the number of commissioners from three to five, with the additional two to be nominated by the OMEA.[48] Drury did not act on either recommendation.

Although Drury wanted the HEPC chairman in cabinet, the knowledge that Beck continued to be popular with the public, the OMEA, and the ILP explains why he did not remove him. However, it was surprising that Drury did not act to control Beck through his government's power to appoint the two other commissioners. He did make one appointment in 1919, Lieutenant-Colonel Dougall Carmichael, a minister without portfolio, but this was only to meet the requirement for one or two cabinet ministers to sit as commissioners. Moreover, this appointment only filled the vacancy created by the death of the third commissioner, McNaught, in February of that year. Thus not only did Lucas, the attorney general in the Hearst government, not resign as the second commissioner, but Drury did not remove him. Ironically, Carmichael had personally run against and defeated Lucas in the election and then urged Drury to keep Lucas on the HEPC. By not appointing a second UFO loyalist, Drury left Beck with the potential to control two of the three commission votes, indicating that he had not initially resolved to restrain Beck.

Beck's operation of the HEPC was nonetheless anathema to Drury's sensibilities. It left the two destined for conflict because the premier had, in Nelles's words, an 'intuitive scepticism of big government, debt, and bureaucracy.' Drury soon became alarmed when he discovered the extent of the financial commitments that were necessary to complete both the Queenston project, already under construction, and the extensive radial projects ready to be initiated. And this was only part of the problem. He was informed shortly afterwards that the projected cost for Queenston was quadruple the original estimate, that Beck had spent $3.5 million more than he had in appropriations, and, worse still, that Beck had padded estimates to generate funds for projects that lacked authorization. As for radials, Beck had proceeded with some construction on the basis of mere promises of authorization from Hearst prior to the election. These findings led Drury to challenge Beck's stewardship of the HEPC because Beck had subverted the cabinet's power to control taxation and spending, central tenets of responsible government which Drury refused to disregard.[49]

Although not yet fully under way, the radial program became the focus of Drury's effort to restrain Beck generally and the government's debt commitments for the HEPC in particular. So Drury appointed Justice R.F. Sutherland of the Supreme Court of Ontario to lead a royal commission investigation of radials on 15 July 1920, and refused to guarantee further radial bonds until Sutherland reported. Drury had initiated the investigation primarily to substantiate his belief that the proposed radials would hurt the credit of the government and not be viable alongside the government's own plans for a network of provincial highways. As part of the UFO's election platform, these roads were designed to meet the recent and rapid growth in automobile use, but Drury had an additional reason for arriving at his decision. The minister of public works, F.C. Biggs, the Adam Beck of highways, had led a campaign against radials with the aid of his own powerful supporting cast of interest groups.[50]

While the Sutherland Commission was investigating radials, Beck announced on 5 December 1920 that an agreement had been reached on the purchase of the Toronto Power Company's properties. Dubbed the clean-up deal, it was intended to transfer the company's generation, transmission, distribution, and street railway companies into public hands for $32.7 million. The announcement, in Plewman's view, represented the 'surrender ... of the citadel of private power interests in Ontario.' When the transaction took place, the deal was segmented. In May of 1921 statutory approval was granted only to the City of Toronto to purchase the electric distribution and street railway properties within city limits.[51] The remainder of the deal would not be completed until 1924.

Beck undoubtedly held up the remaining elements of the clean-up deal in order to gain leverage in his struggle with Drury over the future of radials. Moreover, the 1921 segment of the deal served his grandiose plans for radials by permitting the City of Toronto to transfer rights of way to the HEPC for east and west radial entrances. Drury, for his part, sought further leverage on the issue of radials when he finally removed Lucas from the commission in July 1921 and replaced him with Fred Miller. Miller had not only pledged to vote with Carmichael against Beck on the use of HEPC funds for municipal radial referendums; he was also a member of the Toronto Transit Commission, a body aligned against Beck's ambitious plans for the radial entrances. Beck responded to Drury's action by naming the deposed Lucas as the HEPC's general solicitor.

Drury's position on radials was bolstered when the Sutherland Commission reported on 31 July 1921. Its main points were that radials would never be self-supporting, were too costly to be undertaken at the same time as the Queenston-Chippawa project, and should not be allowed to proceed until the merits of the provincial highway system could be gauged. In general, it concluded that radials were not an 'essential public utility of real necessity.' The report's impact was blunted shortly after, however, when a minority report by Fred Bancroft, a senior ILP MPP, expressed the opposite sentiment. Despite the implications for the UFO's coalition with the ILP, Drury nevertheless vowed on 24 August to repeal the provisions of the radial act which delegated executive decisions to the HEPC and committed the government to guarantee radial bonds. Not to be outdone, Beck instructed the municipalities to proceed with their radial plans and condemned the conclusions of the Sutherland report in a published counter report on 10 February 1922.[52] He could afford to be so defiant at this time because the Queenston generating station had begun producing power on 29 December to great fanfare.

The reprieve Beck enjoyed following the completion of the Queenston station was interrupted by the release of another report on 10 February 1922, one which compared the HEPC unfavourably with American privately owned and publicly regulated utilities. Although this report, written by consulting engineer W.S. Murray, praised Beck as a man of high esteem and the HEPC's accomplishments as remarkable, it argued that the commission would never have been created if the privately owned utilities in Ontario had been regulated as extensively as they were in some jurisdictions in the United States. Moreover, with the then current rates in Ontario being comparable with those in California, the best example of regulation, it concluded there was no advantage to public ownership. The HEPC responded to this challenge with a published refutation that claimed the Murray report's findings were worthless and would fall into disrepute like the earlier American critiques.[53]

Although publication of the Murray report was a blow to Beck, his reprieve was not brought to an end until 3 March 1922. On that day Carmichael, the UFO cabinet minister on the HEPC, tendered his resignation to Drury because he was unable to explain the discrepancy between the commission's 1921 estimates and the costs for Queenston in that year. The offer, which was rejected, was undoubtedly made for political effect; it had first been made and rejected privately in the fall of 1921. Moreover, Drury had chided Beck early in 1922 for submitting estimates unseen by the two UFO commissioners and thus had known of the problems. Nonetheless, Carmichael's offer served Drury admirably as a vehicle to draw public attention to the huge cost overruns for the Queenston-Chippawa project. It also served as his pretext for appointing another royal commission on 13 April. Chaired by W.D. Gregory, it was established as a full-scale inquiry into the HEPC's operation, not just the Queenston project.[54]

With the HEPC on the defensive, the expected showdown between Drury and Beck over the future of radials did not materialize. Instead, their conflict was resolved through a compromise in June 1922. In essence, Beck agreed to approve the remaining radial portions of the clean-up deal, and Drury agreed to permit part of the radial program to continue. For existing radials, including those transferred to the City of Toronto and the HEPC as part of the clean-up deal, the provisions of the Hydro-Electric Railway Act of 1914 continued to apply. For all new and contemplated radials, however, Drury's Municipal Electric Railway Act of 1922 supplanted the 1914 legislation. It closely followed the recommendations of the Sutherland report, especially the one that the government should not guarantee radial debts, and required new municipal votes for lines to proceed. Although the Toronto to St Catharines line was not revived, a dispute over funds already spent would reach the Judicial Committee of the Privy Council in London in 1929 with implications regarding who owned the HEPC.[55]

The turn of events for Beck's hydro empire under the Drury government did not augur well for its continued autonomy from government. Beck countered this development by engineering a resolution at an OMEA conference in Hamilton on 4 April 1923. It was designed to gain the HEPC's independence from the government, offering 'to take such steps as will relieve the Provincial Government from all financial responsibility, and place the administration of the Hydro-Electric System in the hands of the municipalities who are the owners and are financially responsible for the undertaking.'[56] The resolution was the culmination of Beck's numerous rearguard retorts to Drury's challenges and was likely made with the knowledge that an election was imminent. In sum, Beck had argued that the HEPC was the trustee of a municipal cooperative and thus not subject to the restraints on a government department or crown corporation. Drury, for his part, had argued equally vociferously that responsible government dictated that he exercise control over the HEPC. This conflict continued into the June 1923 election with Beck working to defeat the government and Drury defending the propriety of his actions.

Although Drury had effectively suppressed Beck's radial program, it was a small consolation that did not serve his party well in the election. The platform of the Conservatives under Howard Ferguson, the party leader since 1920, held that the UFO's war on the radials represented a 'repudiation of the principle of public power itself and a denial of cheap transportation to the workingman.' The charge was effective because it was accompanied by a renewal of the HEPC's close ties with the Conservatives through the return of Beck to the party's fold as a candidate. The irony of Beck running against the government while serving at its pleasure did not escape Drury, but he knew there was electoral danger in campaigning against Beck and the HEPC. He nevertheless managed to incur the wrath of municipal hydro leaders by innocuously suggesting that they would not be able to run the HEPC without financial assistance from the government. With his statement interpreted to mean that the commission was not owned by the municipalities,[57] the municipal cooperative myth, revived by the OMEA's resolution two months before the election, contributed to the defeat of the government. The election gave the Conservatives a large majority with seventy-five seats to seventeen UFO, fourteen Liberals, and four ILP.

The outcome of the election was not the blessing Beck might have presumed. Although he had refused to be included in the Hearst and Drury cabinets, he was not in a position to refuse Ferguson's call in 1923. The new premier was not prepared to be bullied by Beck. Ferguson, furthermore, was aided by the fact that Beck was now sixty-six years old, and had begun to suffer the ill-health that would continue until his death in 1925. The election also placed at Ferguson's disposal two positions for commissioners. Carmichael resigned from the HEPC following the election, and J.G. Ramsden, a former Toronto city councillor who had been appointed in January 1923 following the death of Miller, was removed by order-in-council after refusing to resign. Ramsden, Plewman reports, was a machine Liberal who had been very critical of Beck.[58] Ferguson filled one vacancy in July 1923. Returning to the pre-1914 arrangement of having two ministers on the HEPC, he appointed J.R. Cooke who, like Beck, was made a minister without portfolio. Cooke had refined his knowledge of HEPC affairs in the previous legislature through a committee on rural electrification. Given Beck's incapacity, Cooke became the de facto chairman of the HEPC and ran it alone because a third commissioner would not be appointed until 1925.

Although the Gregory Commission had submitted interim reports to the Drury government, the release of the final report was held over until after the 1923 election by mutual consent of the various party leaders. Once in power, Ferguson amended the commission's terms of reference by order-in-council on 12 September in an effort to restrict its scope. The rider following the eighth term of reference had originally asked the commission for 'suggestions and recommendations' and to 'report on the evidence and the facts.' The amendment deleted the former and kept the latter, suggesting

Ferguson wanted to bar the commission from making recommendations.[59] The final report, submitted in March 1924, disregarded Ferguson's instruction. Although especially critical of Beck's apparent contempt for the controls of the legislature and his own statutory duties, the report actually did not recommend that the HEPC be made a department of government as might have been expected.

The Gregory report supported Beck's long-held sentiment and stated that exposing the HEPC to patronage would be its 'fatal blow.' But it offered the seemingly contradictory recommendation that the HEPC should be both accountable to the government and the legislature *and* enjoy 'full liberty within itself.' This position rested upon the argument that what the government lacked in direct control over the HEPC it could recoup through the indirect control it held through appointments.[60] The important corollary was that government was responsible if and when commissioners were not removed for wrongdoing. This view suggested that Drury was at fault for not removing Beck and for waiting so long to remove Lucas, but it gave no recognition to the precariousness of his minority government. There was an additional corollary, however; the suitability of the crown corporation structure of the HEPC which had developed at Beck's urging was affirmed.

Although the Gregory report thereby supported Beck's view of corporate autonomy, it strongly criticized the contention continually made by Beck and the OMEA that the HEPC was the trustee of a municipal cooperative. This claim was said to be 'misleading and inaccurate.' The report came to this view by critically testing the OMEA resolution adopted in April 1923 which had called on the government to transfer formal ownership of the HEPC to the municipal hydro commissions. The resolution was not well grounded, according to the report, because it lacked explicit statutory support in two regards. First, the Power Commission Act was said to be a 'rejection of the principles' underlying the municipal cooperative foundation of the Ross Power Act. Secondly, although the Power Commission Act provided for sinking fund payments by the municipalities to retire the HEPC's debt, there was no direct responsibility on them to ensure the debt was fully retired. Moreover, the report maintained that the sinking fund payments only meant the municipal hydro commissions would become the owners if and when they had relieved the province of the HEPC's debt. The report found this to be an unlikely scenario because the contributions to the fund by 1924 totalled only 2 per cent of the debt and were not 'materially increasing.' The report concluded, therefore, that the OMEA resolution, and with it the municipal cooperative claim, was impractical.[61]

Despite its conclusions supporting the view that the HEPC needed independence from both the government and the municipalities, the Gregory report's recommendations were largely devoted to improving weaknesses identified in the commission's operation. Its specific concerns centred on the exercise of business judgment, especially on decisions affecting the cost of power. The report's criticisms on this score, however, were muted by its resounding tribute to Beck's accomplishments. Having questioned his 'lack of frankness' and his 'arbitrary and inconsiderate' behaviour, it nonetheless suggested that he had 'rendered most notable service' and deserved 'full and ungrudging credit' for the HEPC's success.[62]

The report could be critical of the business management of the HEPC and Beck's character while also paying tribute to him because it distinguished between his methods and objectives. By the time the report appeared, the commission had acquired ninety private companies through a strategy primarily guided by rate competition. The report found fault where this strategy involved methods that were without a statutory foundation, but in the end it endorsed the principle of public ownership. In fact, it stated: 'The principle of public ownership of the water powers of the Province and their development by the people for the people, is, in our opinion, fundamentally sound and should be maintained at all hazards in its full integrity.' The report, therefore, provided a *post hoc* justification for the fact that Beck had transformed the HEPC from its original state of co-existence with private enterprise to, for all intents and purposes, a public ownership monopoly. Notwithstanding this endorsement, Beck issued a published critique of the report.[63]

On the issue of salary compensation for the commissioners, the Gregory report supported the view that the HEPC should be treated as a crown corporation. It recommended that their salaries should be charged to the operation of the commission rather than be paid, as part had been, by the government. As a result, the government amended the Power Commission Act in 1924 to leave salary compensation to be raised in full through a municipal levy. Thus, there was an irony to the implementation of this part of the salary recommendation. The report had belittled the notion that the HEPC was the trustee of a municipal cooperative, yet the government now enhanced this aspect of its nature. In addition to amending the Power Commission Act to relieve the government of paying $6,000 towards Beck's salary, as had been the case since 1912, the act was also amended to permit the combined salaries of the commissioners to be increased to $45,000 from $15,000. This was a figure the Gregory Commission would have supported. Although it had uncovered that the HEPC commissioners and senior officers were drawing large salaries out of the Ontario Power Company, which the HEPC had operated as a joint-stock subsidiary since 1917, its report objected only to the surreptitious way in which salaries were financed rather than to the overall level of compensation.[64]

Closely following the release of the Gregory report came statutory approval for the unfinished elements of the clean-up deal. An agreement for the transfer of the remaining assets of the Toronto Power Company to the HEPC was reached on 25 March 1924. The act also provided for the consolidation of the Ontario Power Company into the commission but did not do the same for the Central Ontario Power Company. It remained owned by the government and operated by the HEPC until 1930, even though the Gregory report had recommended that both subsidiary companies be consolidated into the commission.[65] The HEPC had nonetheless become, for all intents and purposes, a public ownership monopoly. With Beck's death on 15 August 1925, one year after this milestone, the HEPC entered a new era, having to define a new mission for itself in the absence of the personal leadership that had so far provided its bearings.

4

Injecting the Hydro-Electric Power Commission with Political Sensitivity: The Ferguson Formula, 1925–43

From the death of Adam Beck in 1925 to the watershed provincial election in 1943, the government continually struggled to fill the void of his personal leadership of the Hydro-Electric Power Commission. This new emphasis on appointments reflected a change in the relative importance the government had previously placed on its two instruments for setting the direction of the HEPC. Statutory enactments had been used extensively in the Beck era, but their drawback was the attendant political responsibility that came with this degree of formal control. Appointments, on the other hand, when used strategically and judiciously, were found to bypass this burden in the post-Beck era. Rather than use them to increase government control, as might have been expected, the government used them to provide itself and the Ontario Municipal Electric Association with informal influence in HEPC affairs. In the process, the government institutionalized further the seemingly ambivalent nature of the commission as both government corporation and trustee of a municipal cooperative. This character of appointments developed over three distinct periods of the post-Beck era.

The dominant theme of the first period, the late 1920s, was a federal-provincial jurisdictional dispute. Given its intentions to control power developments on the Ottawa and St Lawrence rivers, the federal government posed a potential threat to the HEPC's sector-wide existence as a first-generation crown corporation. In the face of this challenge, Premier Ferguson appointed a chairman from outside the government and the commission. He also gave the OMEA representation on the commission symbolically equal to what the government received by statute. With a minister no longer serving as chairman, the government could distance itself from political responsibility for the commission's affairs while ensuring continued influence. The Ferguson formula had the further advantage of uniting the previously disparate views on whether the HEPC's nature around the need not to cede jurisdiction to Ottawa. By resisting the federal incursion, the commission was transformed into a nationalistic crown corporation.

In the second period, marked by the political turbulence and partisan wrangling of the depression, neither George Henry nor Mitch Hepburn followed the Ferguson formula. The upshot was that in loading the HEPC with partisans, neither premier was able to dissociate his government from political responsibility for the commission's problems. Henry had to accept responsibility for the HEPC's failings after it amassed a huge surplus from contracts signed with companies wholly within Quebec. This had been a nationalistic response to the earlier federal jurisdictional challenge. Hepburn's use of enactments in addition to appointments for his own nationalistic policy of repudiating the Quebec contracts also left him little distance from the HEPC's problems.

The economic renewal and political stability of the late 1930s and early years of the Second World War set the parameters for the third period. Reuniting the government and the OMEA in common cause, Hepburn returned to the safety of the Ferguson formula in 1937, elevating it to the status of a 'semi-convention.' His action proved to be important for war mobilization, one outcome of which was an agreement to develop the Ottawa River. Although in this agreement a degree of federal jurisdiction was accepted, the HEPC nonetheless became unrivalled as a first-generation crown corporation as a result. Thus, by the 1943 election the commission had let its earlier nationalistic stridency pass to assume the role of a facilitative corporation.

Expansion in the Roaring Twenties

Premier Ferguson's first task following the death of Adam Beck was to find a new chairman. Ferguson had in mind a person who would exercise the HEPC's business autonomy in a manner that was politically sensitive to the needs of the government. This was important to Ferguson who had lectured Beck that 'it [was] not only embarrassing but humiliating ... to be placed in the position where I am forced to say that I know nothing of some project for which the Government must take responsibility.' On 12 September 1925 Ferguson appointed Charles Magrath, a non-partisan from outside the government and the HEPC. Magrath's past as a federal official with diplomatic experience yielded, Denison writes, a chairman who was as restrained and self-effacing as Beck had been flamboyant. Moreover, Magrath's focus on legal, fiscal, and financial matters revealed that the crusade of Beck's personal leadership was over.[1] His incumbency demonstrated that the power to appoint could be the power to remake the HEPC.

Ferguson's desire not to be manipulated by the commission or its chairman was also evident in his selection of the two other commissioners. He used these appointments to remake the HEPC's relationship with the government and municipalities by creating constituency-like representation. As a government representative, J.R. Cooke remained a commissioner, filling the statutory requirement for one or two cabinet ministers. The third commissioner, appointed with Magrath on 12 September, was Alfred Maguire, president of the OMEA since 1923. He was also a well-known Liberal who had supported Beck in the radial dispute with Drury while mayor of Toronto. In appointing Maguire, Ferguson was responding to the OMEA's persistent call for representation on the HEPC. Rather than give the association direct representation, Ferguson asked for a nominee. Maguire, who was the nominee, later stated that he was appointed, without condition, to be a representative of the OMEA.[2]

Magrath had developed a certain expertise on water power matters. Since 1914 he had been chairman of the Canadian section of the International Joint Commission, the body which managed boundary water issues between Canada and the United States, and he continued in this post while HEPC chairman. His knowledge was specifically needed to guide the HEPC through the jurisdictional quagmire that had developed by this time over southern Ontario's other large and accessible water power sites, the St Lawrence and Ottawa rivers. Their power was needed if a shortage was to be averted in the late 1920s. The Ottawa was caught in a federal-provincial jurisdictional dispute and the St Lawrence was further complicated by the need for a Canada–United States agreement. Despite the greater jurisdictional problems, the St Lawrence was more attractive to the HEPC for development than the Ottawa, or for that matter than power purchases from private companies inside Quebec. Its proximity to southwestern Ontario meant lower transmission costs. In addition, water power continued to be considered preferable to thermal power generation with American coal because of low operating costs.[3]

Underlying the federal-provincial dispute was the federal government's claim to control water power 'incidental' to its constitutional jurisdiction over navigation and shipping. This contention was augmented by the reference to water power on canals in the third schedule of the Constitution Act, 1867 dealing with federal public works and property. For its part, since an

1898 decision of the Judicial Committee of the Privy Council, Ontario had believed that it had a 'proprietary right' to water power on navigable rivers. The Judicial Committee, however, had not determined the respective limits of the two jurisdictions.[4] This meant the federal-provincial dispute had two broad implications: it complicated the negotiation of a Canada–United States agreement on the St Lawrence, whose international section was clearly outside provincial jurisdiction; and it brought the legality of past and prospective HEPC developments on domestic navigable rivers into question.

The HEPC's campaign to develop power on the St Lawrence, although first begun at the close of the First World War, was in limbo when Magrath became chairman. In May 1924, as a result of opposition from vested transportation and electric power interests in Montreal, Prime Minister Mackenzie King had appointed an extra-parliamentary national advisory committee on the development of the river and notified the International Joint Commission that Canada favoured further study.[5] As a result of this dalliance, Ferguson had explored Ontario's higher-cost water power alternatives with Quebec premier Alexandre Taschereau in January 1925. The premiers condemned the federal government for having previously granted water rights on the Ottawa River, arguing that they were in provincial jurisdiction, and agreed to terms for the HEPC to purchase power from Quebec companies. Although Taschereau had previously opposed such a transaction, he was persuaded to soften his position. Ferguson, who was also minister of education, offered him the *quid pro quo* of a commission to investigate the problem of French-language schools in Ontario, a perennial thorn in Quebec-Ontario relations. Taschereau, for his part, agreed to the export of power if provincial jurisdictional claims on the Ottawa River failed.[6]

The federal-provincial dispute on the Ottawa was exacerbated when, in the knowledge that the HEPC was in urgent need of power, competing private applications were made for federal government leases on the river in 1925 and 1926. The first came in May 1925 when a revised lease for Carillon Falls, between Ottawa and Montreal, was tabled in the House of Commons. The application had been made by the National Hydro-Electric Company, recently acquired by Dr Wilfrid Laurier McDougald. Not only was he a prominent Quebec Liberal, but King had appointed him chairman of the Montreal Harbour Commission in 1922 and a member of King's national advisory committee in 1924. Both roles were significant for the HEPC. The former position allied him against power developments on the St Lawrence because the accompanying canals would impinge on the port of Montreal. The latter meant he was in a position to frustrate the HEPC's efforts to develop the St Lawrence, leaving the commission little alternative but to buy power from him on the Ottawa River. The second application was made in the spring of 1926 by the sons of Sir Clifford Sifton, himself a prominent Liberal and a member of the advisory committee. Having acquired the company in late 1925, they sought to renew a federal charter for the long-dormant Montreal, Ottawa, and Georgian Bay Canal Company. They wanted control not only of the Ottawa River's entire potential; the original charter, granted in 1894, had explicitly included the water power at Carillon Falls.[7]

Although both companies offered to sell it power, the HEPC refused to prejudice Ontario's claims to control power developments on the Ottawa by dealing with either. Instead, the commission met its immediate power needs by contracting with the Gatineau Power Company, wholly within Quebec, on 19 May 1926. Although this purchase signalled a break with the HEPC's past self-reliance, under the circumstances it was endorsed by the executive of the OMEA, of which Maguire was still president. Meanwhile, the battle for control of the Carillon site continued, but with a new player. The Shawinigan Power Company, owned by Sir Herbert Holt, a prominent Conservative, took control of McDougald's company and the associated application for a revised lease in June.[8]

The fate of Carillon Falls was tangled in the two changes of government in Ottawa in the summer of 1926. After failing to receive a dissolution, King resigned as prime minister on 28 June and Governor General Byng installed Conservative leader Arthur Meighen in his place without an election. Notably, on 25 June King had advised that McDougald be appointed to the Senate, but Byng refused. With the decision on the Sifton and Holt applications now in Meighen's hands, Holt was granted the lease in August, after the tenuous Conservative government itself had fallen. When the Liberals won the 14 September election, King thereupon cancelled the lease, believing it was granted to Holt in exchange for a campaign contribution. In addition, King also had McDougald appointed to the Senate. So as not to play favourites, King did allow Holt's original lease to be temporarily extended to 1 May 1927, which was the same day the Siftons' charter would face renewal.[9] Given the circumstances, however, the chances of Holt's application being revived were remote.

At this point, Ferguson and Magrath orchestrated a campaign to defeat the Siftons' charter renewal, and with it federal jurisdictional claims on the Ottawa and St Lawrence rivers generally. Although their motivations were complementary, they were not identical. Whereas Ferguson was interested in protecting Ontario's jurisdictional claims, Magrath was also promoting the integrity of the HEPC as a public ownership monopoly. This became evident on 11 November 1926 when the government, by order-in-council, leased the water rights on the Abitibi Canyon to the Ontario Power Service Corporation (OPSC), a subsidiary of the Abitibi Power and Paper Company, without consulting the HEPC. According to Nelles, Magrath favoured a rational development of northern power resources by the commission rather than uncoordinated private development. Ferguson, however, felt this would be an unnecessary subsidy to mining and paper companies. In any event, the OPSC would have to sell power to the HEPC, whose needs were 250 miles south in Sudbury through virgin territory, for its plans to be feasible.[10]

As for the campaign against the Siftons, Magrath had Ontario's legal arguments vis-à-vis Ottawa refined in February 1927 by lawyer Loring Christie, a former senior external affairs official in the Borden government.[11] For his part, Ferguson initiated a resolution in the legislature condemning the Siftons and claiming interprovincial waters for the provinces. This was passed unanimously on 7 March. The HEPC and the OMEA followed by inundating the federal government and members of Parliament with various forms of public opposition to the charter. Wary of the Siftons' influence, King did allow their charter to be tabled in the House of Commons, but only through a private member's bill. When the bill reached the committee stage, the Liberal majority, on instruction from the cabinet, voted not to report. This meant the Siftons' charter expired on 1 May along with the Holt lease.[12]

Having met the Sifton and Holt challenges, Ferguson sought to expedite the development of Carillon Falls by the HEPC. Although he wanted to negotiate a resolution of the jurisdictional issues, King would not concede Ferguson's view that the division of jurisdiction neatly allocated power developments to the provinces and navigation works to the federal government. King felt that he had a strong case in law for licensing power developments and charging the cost of navigation works against them and thus no need for negotiation. In response, Ferguson and Taschereau recommended at a November 1927 federal-provincial conference that the dispute be referred to the courts. The federal cabinet decided originally to refer seven questions to the Supreme Court of Canada on 21 January 1928. When Ferguson protested provincial exclusion from their formation, it agreed to append an additional two. Ferguson nevertheless believed that the vagueness of the questions would not bring a definitive resolution to the jurisdictional matters.[13]

Even as the federal-provincial jurisdictional dispute was before the Supreme Court, it was exacerbated by two important developments. The first followed from the application of the Beauharnois Light, Heat and Power Company on 17 January for federal cabinet approval to develop the Quebec section of the St Lawrence. In response to this application, King's national advisory committee on the St Lawrence, which had not met in three years, assembled later in the month after King requested a report. It recommended that the Quebec section of the river be developed ahead of the international section adjacent to Ontario and that the HEPC meet its power needs from Beauharnois. The recommendation reflected well-known conflicts of interest. Senator McDougald was financially interested in Beauharnois and Senator Sifton's son, Winfield, was its Ottawa lobbyist.[14]

The second development was the lease Beauharnois received from the Taschereau government for the water rights to the Quebec section of the St Lawrence on 23 June. Although the lease appeared to upstage the Supreme Court's decision on jurisdiction, still to be rendered, King did not object. The Quebec government had agreed that the lease would not prejudice the federal government's proprietary rights if the court determined that the federal government owned the water.[15] Such an agreement was precisely what Ferguson had sought from King, but Ontario did not have Quebec's capacity to act unilaterally because of the international implications of its proposal. The consequence of these two developments was that the HEPC would have to meet its future power needs on the Ottawa River or from Beauharnois.

The Supreme Court's decision on navigation/power development issues, delivered 5 February 1929, was a victory for the provinces. Written by Justice Lyman Duff, it stated that the federal government had no right to the 'whole beneficial interest' of navigable rivers and, therefore, could not constitutionally assume control over water power not needed for navigation. On the basis of this decision, Ferguson sought to negotiate an end to the jurisdictional dispute, but King argued that the decision had not removed the federal government's authority to license power developments. King's position was seen in its full light four weeks later on 8 March when the federal cabinet approved Beauharnois promoter Robert Sweezey's plan to finance and build a dual purpose canal/power development on the Quebec section of the St Lawrence. In approving the project, King conceded the provincial rights position on the ownership of water on navigable rivers, as the court had declared, but he did not relinquish the power to approve such projects or to ensure that power developments financed navigation works.[16] Neither had been a question referred to or resolved by the Supreme Court. As a result, the HEPC could not even proceed on the Ottawa River.

Although Sweezey's victories with the Taschereau and King governments were important, he needed to secure a large power contract with the HEPC to receive financing for the Beauharnois project. Indeed, according to T.D. Regehr, Sweezey had reached an informal understanding with the HEPC before he received the federal cabinet's approval. A tentative agreement with the commission was later settled on 10 June 1929, but its finalization was delayed to permit Ferguson to hold an election first. In the meantime, the OMEA, having been consulted by the HEPC, gave its approval for the power purchase. The commission then signed the tentative agreement with Beauharnois on 28 October, enabling the company to arrange its financing, propitiously, one day before the stock market crash that would precipitate the great depression. After the Conservatives were returned to power on 30 October, the Ontario cabinet approved the agreement on 29 November, with the official contract signed the following day.[17] In the intervening month, the government had presumably assessed the possible implications of the economic downturn for power demand before approving the purchase.

Given the Ontario government's open involvement in the HEPC's affairs, members of the legislature had questioned the dealings with Quebec companies in early 1929. According to Nelles, Magrath refused to provide any information on the grounds that business secrecy was as imperative in the public sector as in the private. In Magrath's view, 'disclosure should be left to the discretion of the commission ... and if the Legislature questioned its judgment, and after the pros and cons had been fully debated and good cause established, then the commission would submit to its will. But on "sound business lines" ... the government should protect the commission against the meddling interference of the Legislature.'[18] Ferguson supported Magrath's position, and both justified themselves on the grounds that the HEPC should not have to submit to the same degree of scrutiny as a government department.

As future events revealed, Ferguson's and Magrath's view on disclosure would prove to be sustainable only when the economy was buoyant. It would not be sufficient to shield the HEPC from criticism of the contracts with Gatineau and Beauharnois and of others made with the Ottawa Valley Power Company in February of 1930 and 1931, with the James MacLaren Power Company in December 1930 and January 1931, and a contract with the OPSC in northern Ontario in April 1930. Moreover, Ferguson and Magrath would resign in 1930 and 1931 respectively and henceforth would not be in a position to defend their view of the HEPC's accountability and the rationale for the contracts.

The HEPC's ambivalent ownership also surfaced as an issue in 1929 through a Judicial Committee decision. When construction of the Toronto–St Catharines radial railway was terminated by the Drury government's radial act of 1922, the financial contributions of a number of municipalities were lost and the commission was still owed funds for the uncompleted project. In response, St Catharines launched a court action against the HEPC for recovery, to which the commission made a counterclaim for St Catharines' full share of the costs incurred. The basis for St Catharines' claim was that its contract with the HEPC ought properly to be considered repudiated by the government, with the commission liable, because the utility was a 'department of government.' For its case, the HEPC argued that, because of its statutory existence, it was an independent legal entity, one which simply had government-appointed commissioners. The Judicial Committee accepted the latter view, giving legal confirmation to the corporate autonomy Beck had promoted.[19]

Partisan Wrangling in the Depression

The years 1930 and 1931 brought a number of important changes to the composition of the leadership of the HEPC and Ontario political parties. After seven years as premier, Ferguson stepped down at the relatively young age of fifty-five on 15 December 1930. He was succeeded by the minister of public works and highways, George Henry, four years his elder. Mitchell Hepburn, a zealous and outspoken federal MP who was only thirty-four, became provincial Liberal leader two days later, while retaining his seat in the House of Commons. And Magrath, now seventy, retired as HEPC chairman on 6 February 1931 without an immediate successor.

With the Liberals having been out of power since 1905, Hepburn quickly invoked HEPC issues in an effort to reverse his party's fortunes. The political vulnerability of the government had become apparent to the Liberals when the first power surplus of the depression was recorded in early 1931. During the legislative session they questioned the seeming paradox of why the HEPC imported large amounts of electricity from Quebec while it exported equally large amounts from Niagara at a net loss. Although Henry responded that the commission was bound to the export contracts of the private companies it had previously purchased, the logic of this arrangement was lost on the public. Following the session the HEPC, still without a chairman, came to its own defence on 6 May. Under Cooke, the minister without portfolio, and Maguire, still the OMEA president, it criticized the Liberals' source of information, namely the bond prospectuses of the Quebec companies. Hepburn replied by charging that this foray into the partisan arena proved the HEPC had been 'corrupted by politics.'[20]

Although Hepburn did not relent in his criticisms, he made a significant political blunder. In a speech in Milton on 21 May he charged that power was now more expensive to produce in Ontario than it was in the United States, failing to recognize that the source of his figures was controversial in itself. Hepburn had quoted directly from a propaganda pamphlet published by a Chicago-based private power lobby dedicated to undermining the HEPC.[21] This incident, which would be repeatedly used against Hepburn, undoubtedly confirmed for the government that poor judgment would be Hepburn's undoing. By adopting this outlook, Henry would misjudge the adversarial temperament of the revived Liberal opposition.

In his appointments to the HEPC on 8 June, Henry abandoned Ferguson's formula of appointing a non-partisan chairman alongside government and municipal constituency representatives. Instead, he stacked the commission with partisans. Alongside Maguire as the OMEA representative, Cooke was elevated to chairman while remaining a minister without portfolio, and former Conservative prime minister Arthur Meighen was made a commissioner even though he did not represent a HEPC constituency. To make matters worse, at the end of a widely reported speech to the OMEA summer convention on 26 June, the HEPC's chief engineer, Frederick Gaby, castigated Hepburn for gross exaggeration in his criticisms of the HEPC. Gaby also did not fail to point out to these disciples of public ownership that Hepburn's source of information had been a private power lobby.[22] As a result of this partisanship, the HEPC was drawn further into the warfare of government and opposition by its own commissioners and staff.

Hepburn soon found a new reason to criticize the Henry government and the HEPC. During the summer of 1931 a House of Commons special select committee investigating the Beauharnois power development uncovered a political scandal that involved the commission. During the committee's hearings, Sweezey, the Beauharnois promoter, testified that he had given the Ontario Conservative party $125,000 through John Aird Jr in exchange for the HEPC's contract with Beauharnois. When questioned, Aird, who was a consulting engineer by profession, stated that the money was for engineering services rendered, a view which Sweezey denied. The committee, however, failed to establish a definitive connection between Aird and the Ontario Conservative party. Aird, whose father was president of the Bank of Commerce, had presented the committee with a letter from his bankers stating that the bonds were still in his possession.[23]

The Hepburn Liberals nonetheless argued throughout the fall of 1931 that Ontario taxpayers were left financially responsible for the mismanagement of the HEPC, now adding to the list of commission faults the contract with the Ontario Power Service Corporation. In a thinly veiled counter-attack, Maguire addressed the charges at the OMEA convention in January 1932. He stated that he had not seen any evidence of interference by government in the HEPC's policy or administration and that the criticism was inspired by opponents of public ownership.[24]

In an effort to clear the air of Hepburn's various charges, on 3 February Henry appointed a royal commission under Justice W.E. Middleton but did not include in its terms of reference the thorny issue of the Beauharnois payment to Aird.[25] Despite Henry's efforts, the Beauharnois scandal remained prominent. On the same day that Henry appointed the inquiry, Prime Minister R.B. Bennett appointed HEPC commissioner Meighen to the post of government leader in the Senate and minister without portfolio. Meighen's first assignment was to lead a Senate select committee inquiry into the improprieties of three Liberal senators who had been implicated in the Beauharnois scandal – McDougald, Andrew Haydon, and Donat Raymond.

Meanwhile, with Hepburn's criticisms of the HEPC and the government continuing, Henry sought to silence him. In a speech in the legislature on 18 February, the premier repeated the charge that Hepburn's speeches were inspired by anti-public ownership sentiment in the United States. Although devastating for a time, the impact of this counter-attack was blunted when Senator Haydon testified before Meighen's Senate inquiry on 15 March. He claimed that Ferguson had personally received $200,000 from Sweezey in addition to the $125,000 paid to Aird. Although the charge was not substantiated, Justice Middleton recommended that the mandate of his royal commission be extended to include the payment to Aird, and Henry agreed in order to placate the Ontario opposition.[26]

The creation of the royal commission nonetheless provided the HEPC and the government with a reprieve from criticism, one which allowed Henry to deal with another potentially explosive issue which had yet to surface publicly. In February 1932 he had been informed that the OPSC was in danger of financial collapse with its generating station uncompleted. Although Ferguson had initially encouraged him to add the company's lease to the royal commission's terms of reference, Henry, who was incapacitated by an ulcer at the time, held the issue over until after the legislative session had ended. On 25 June, with the company facing receivership on 1 July, Henry announced that the government would negotiate for its purchase. A debt-equity exchange was arranged on 28 July, with $18 million in HEPC debentures to be traded for $20 million in OPSC bonds. After resolving this issue, the government had only to worry about the royal commission, which reported 31 October. It exonerated the HEPC of any wrongdoing, but the public, according to John Saywell, remained 'suspicious.'[27]

Now cautious not to criticize Hepburn or incite the Liberals generally, Gaby presented a reasoned explanation for the HEPC's power surplus in a speech to the OMEA convention in January 1933. His general point was that the commission did not have an indication of the severity or prolonged character of the depression when it had planned supply for the 1930s. In purchasing Quebec power, it had decided to increase supply by 7.7 per cent, a figure which was 30 per cent below its historical 11.4 per cent average growth in demand. Although this still resulted in a larger than normal power surplus, or reserve, Gaby added that the HEPC would have been remiss if it had not provided for post-depression recovery; a utility cannot wait for demand to materialize before making provisions for supply. In sum, he claimed that the commission had planned supply with 'reasonable caution' and that the reserves would be justified when the economy improved.[28]

This moral high ground was soon lost when the government was rocked by a scandal over the HEPC's purchase of the OPSC. On 22 March the Progressive leader, Harry Nixon, who was in alliance with the Liberals, asked Henry in the legislature whether any member of the cabinet or any of the commissioners had a financial stake in the company at the time of its purchase. With rumours of impropriety circulating, the question was loaded because Henry had personally negotiated the deal at only a 10 per cent discount, and the commissioners had given it their formal approval. Two weeks passed before Henry responded on 5 April that he had owned $25,000 of OPSC bonds since August 1930 and that Meighen had owned $3,000 since November 1931. On another question Henry responded that Meighen, as chairman of Canadian General Investments Limited, had controlled OPSC bonds since 1929. Henry added, however, that because Meighen had not been privy to the negotiations for the company, it was not necessary to disclose his company's investments.[29]

Seeking to elevate the scandal to the level of Beauharnois, Hepburn deplored the use of public funds to save Henry's and Meighen's investments. Henry, for his part, apologized for the impropriety. According to Neil McKenty, the premier was able to survive the ordeal because it was his political judgment rather than his honesty that was in question. Meighen, for his part, rose in the Senate on 6 April 1933, the day after Henry's revelation, to state that he had done nothing wrong. He nonetheless faced lingering criticism that he had profited from prior knowledge of what price would be paid for the OPSC bonds. In his view, the criticism had been orchestrated to make his indictment of the Liberal senators over Beauharnois appear hypocritical. Although he wanted to counter this revenge, he felt powerless to do so because Henry had refused his requests for an inquiry.[30]

Since this scandal was inhibiting the government from coming to the HEPC's defence, the commission participated directly in a partisan exchange with the Liberals. It began on 19 August when Arthur Roebuck, a nominated Liberal candidate whom Hepburn had publicly promised to appoint attorney general, gave a widely reported speech decrying 'the wreck of the Hydro.' Three days later the HEPC published a pamphlet purporting to 'examine and correct' Hepburn's 'misleading assertions' on its finances and power reserves. In addition, it challenged Hepburn in the name of 'personal probity' to make public retractions. The next day Hepburn responded that the public was less concerned with himself than with the power surplus the HEPC had created to 'further the interests of the Tory party and the Quebec power barons.' Rather than let the matter rest, the commission published a second pamphlet on 31 August. It argued that Hepburn should 'publicly and without ambiguity' withdraw the assertions.[31]

Likely needing to justify the HEPC's counter-attack on Hepburn, Cooke sought to distance the commission of which he was chairman from the government of which he was a minister. In a September speech to a Conservative party conference, he argued that the HEPC was the trustee of a municipal cooperative rather than a government enterprise. He also maintained that the government's supervision of HEPC affairs was limited to the debt it had guaranteed.[32] Whether this separation could be believed or not, the impropriety of the commission's open criticism of Hepburn was not lost on the Liberals. For them, it was confirmation that the HEPC leadership – Cooke, Meighen, Maguire, and Gaby, among others – was as partisan as the Henry government.

With the government needing to go to the polls in 1934, Maguire rekindled the HEPC's offensive for the expected election at the OMEA convention in January. His address reflected his conflicting loyalties as the association's president and a HEPC commissioner, as well as indiscreet partisanship. In addition to arguing that the commission would one day be commended for the foresight of the Quebec contracts, he exhorted OMEA members to be 'loyal to hydro and public ownership.' When reporters asked whether he was accusing the Liberals of being in cahoots with private power interests, he stated on 15 February that the HEPC and the municipalities 'need to be increasingly on guard' against attacks but refused to name any assailants. The message was later made clear when Nixon, in the banter of legislative debate, stated on 13 March that the Liberals, if elected, would fire Gaby for criticizing Hepburn before the OMEA in 1931. For his part, Cooke took up the challenge by defending Gaby's speech as impersonal in nature and the references to Hepburn as 'incidental.'[33]

Shortly after the legislature prorogued on 3 April the HEPC began to figure prominently in the looming election campaign. Aware of the government's election plans, Cooke sent a letter as HEPC chairman to all municipal hydro commissioners on 20 April. He drew to their attention the provincial commission's past difficulty in countering the 'persistent repetition of misstatement' and called on them to 'perform a public service of real value' by distributing and making reference in public discussion to four memoranda he included with the letter. They all focused upon likely election issues. Ten days later, with Hepburn promising to appoint an inquiry into the purchase of the Ontario Power Service Corporation if elected, Meighen resigned as a HEPC commissioner. His reason was that the continuing allegations against him were an embarrassment to the government. Being a federal minister and government leader in the Senate, Meighen then requested that Prime Minister Bennett appoint an inquiry. His intention was to forestall the investigation promised by Hepburn.[34]

Following the election call, the HEPC participated openly in the campaign. It published *Paid-for Propaganda???: Who Instigates Attacks on Hydro?* on 7 June after surreptitiously hiring private investigators to report on the Liberals' connections to the anti-public ownership lobby in the United States. While no damaging information was uncovered, the pamphlet nevertheless claimed that an 'intensified' assault on the HEPC by such private interests was under way. Excusing the lack of specifics, it stated '*secrecy of origin is a characteristic of private-utility propaganda*.' The HEPC hoped the public would be 'aroused to the necessity of resisting the insidious effects' of these efforts so that the 'cooperative municipal undertaking' could continue to serve the province. The Conservatives, for their part, having been in power for all but four years of the HEPC's existence, immodestly claimed during the campaign that the commission's many achievements were their own.[35]

Having entered the partisan fray, both the HEPC and the OMEA left themselves poorly positioned after the Liberal election victory on 19 June. Maguire, who was still OMEA president, recognized this himself when he offered his resignation to the association's summer convention nine days later. Although the offer was not accepted, he nevertheless faced criticism from the delegates for politicizing the HEPC's affairs. The convention did, however, express alarm at the prospect of governmental interference from the Liberals and the threat to fire Gaby. In response, it decided to send a delegation to meet with Hepburn. On the federal level, Senator Meighen's request for an inquiry was granted. Bennett appointed Lyman Duff, now chief justice, on 6 July,[36] just days before Hepburn was to take office. This would not, however, dissuade Hepburn from launching his own inquiry with a broader mandate.

Economic Turmoil over the Quebec Contracts

The HEPC's partisanship in the 1934 election guaranteed a tumultuous relationship after the Liberals took office on 10 July. The next day Hepburn removed the remaining Henry government commissioners, minister without portfolio Cooke and OMEA president Maguire, by order-in-council. For chairman, Hepburn appointed Stewart Lyon, the sixty-eight-year-old editor of the *Globe*. Lyon had been a strong supporter of the HEPC throughout the Beck era, but he had subsequently backed the Liberals' attacks on the Henry government and the commission's management. For commissioners, Hepburn appointed two ministers *with* portfolio, Arthur Roebuck, the attorney general who had been a leading critic of the HEPC, and Thomas McQuesten, the minister of public works and highways. The appointment of two ministers, which was the government's statutory prerogative, meant an overall majority of the commissioners was subject to cabinet control. Thus, although Hepburn had followed the Ferguson formula in part by appointing a non-partisan outsider as chairman rather than selecting the chairman from cabinet, he did not balance government and municipal constituency representation. The OMEA had not only lost its representation; Hepburn also refused to give the association its requested hearing.

The new HEPC commissioners met for the first time on 12 July, the day after their appointment. Since Hepburn had stated immediately after his government had been sworn in that he would insist on Gaby being fired, the commissioners wasted little time before cleaning house. Gaby and I.B. Lucas, the former commissioner and attorney general in the Hearst government, were fired outright. In addition, the forty-six employees with salaries over $5,000 were dismissed, although they had the option of renegotiating their salaries. A.V. White, later revealed as the author of *Paid for Propaganda*, would also be fired. Gaby's former duties were then divided, with Thomas

Hogg promoted to chief engineer, hydraulic and operation, and Richard Jeffery promoted to chief engineer, municipal relations and rural power.[37]

The same day Hepburn appointed his promised royal commission inquiry into the Ontario Power Service Corporation under Justices F.R. Latchford and Robert Smith. Reporting on 20 October, the inquiry concluded that Henry had been precluded from participating with propriety in the purchase of the company because he had a personal financial interest of $25,000 and a previously undisclosed business interest of $200,000 as a director of the North American Life Insurance Company. Moreover, in addition to shedding light on Henry's conflict of interest, the inquiry questioned his business judgment for negotiating the purchase of the OPSC's bonds at 90 per cent of their face value. Although Henry emphasized that the bonds would have yielded only 71 per cent if they had been traded on the open market, the inquiry determined that they were trading at 30 to 40 per cent in June 1932 when he had announced that the government would purchase the company.[38]

Meighen fared worse. Chief Justice Duff had left the country before beginning his inquiry and on his return announced on 16 July 1934 that he would not proceed, given that the Latchford-Smith inquiry was already under way. The Latchford-Smith report determined that in addition to Meighen's personal interest of $3,000 in OPSC bonds, he had controlled another $300,000 in the name of Canadian General Securities. More damaging, however, was another fact. Immediately after Henry formally requested the HEPC to purchase the company on 28 July 1932 at 90 per cent of face value, Meighen had traded extensively in its bonds for as little as 62.5 per cent. Although this action could be defended, that he simultaneously voted as a commissioner to accept the terms of the company's purchase on 2 August without disclosing his financial interest was considered unethical. Meighen had argued that the vote was a *pro forma* ratification of the government's terms, but the inquiry found no statutory basis for this view. In addition to finding that he had an obligation to disclose his interest, it stated that he should have removed himself from the decision.[39]

Although Meighen was vindicated of the charge of insider trading, he was soon after included in a civil law suit launched by the HEPC to recover $4,553 of its funds. The case against him, Maguire, Gaby, and Lucas arose following Lyon's discovery in October 1934 that the commissioners had hired private detectives to spy on the Liberals prior to the election. Cooke was notably absent from the suit because he had died on 13 August while the Latchford-Smith hearings were in progress. He had driven his car into a bridge after he 'went to pieces' under the strain of events, according to Magrath.[40]

On the heels of Lyon's discovery, the new commissioners, two of whom were ministers, took a second action by appointing Murray McCrimmon HEPC controller, a new non-statutory position, on 1 November. Having been assistant secretary since August, he assumed the two posts with the qualification of having held executive positions with the Brazilian Traction, Light, and Power Company. The lateral entry of a person with a private sector background in so sensitive a position was unprecedented for the HEPC, and past association with that particular company undoubtedly caused some discomfort for those wary of the commissioners' intentions. The reason was that William Mackenzie, the former principal of the defunct Toronto Power Company who had done battle with Beck, was Brazilian Traction's founder. In addition, McCrimmon's uncle, Alexander Mackenzie (no relation to William), had been president of the latter and his brother Kenneth was its current corporate secretary.[41]

The HEPC's finances were also the subject of Lyon's address to the January 1935 OMEA convention. He announced that arrangements had been finalized, with statutory changes to follow, for all commission debt to be placed in the name of the HEPC rather than the government, although with the government guarantee continuing. Even though this change was self-serving for the government in that it removed the commission's debt from the public accounts in the midst of the depression, it was an important concession to HEPC autonomy. This did little to appease the delegates. They were still upset by the snub the OMEA had received in Hepburn's appointments to the commission and by his refusal to receive a delegation from the association. In frustration, they unanimously demanded that the government renew their representation. The convention did, however, close on a conciliatory note. President-elect James Simpson, the mayor of Toronto, stated that the OMEA had no right to representation but that the justness of its cause would eventually get it a hearing.[42]

With the lawsuit against the former HEPC leaders ongoing, Maguire, who resigned as OMEA president at the convention, used his last address to defend their use of private detectives to track 'insidious propaganda.' He claimed it was a practice which had been begun by Beck for the sake of 'eternal vigilance.' Although the trial did not commence until 1936, the suit would eventually be dismissed, but not without the judge stating that the HEPC's case had failed miserably. The case had nonetheless permitted Lyon, and presumably Hepburn, to humiliate the commission's former leadership.[43]

As with the experience of the Henry government, Hepburn brought on his own political troubles when his government's control over the HEPC was extended into its business decisions. During the winter of 1935 the government proceeded with its plan to renegotiate the Quebec contracts, having vowed to do so prior to the election and to go 'back to Niagara' for future power requirements.[44] When the legislative session began, Roebuck led the government's attack with a nine-hour speech spread over three days in late February. Backed by research from Jeffery, the chief engineer for municipal relations and rural power (which Jeffery later disavowed), and a legal opinion from constitutional lawyer Lewis Duncan, Roebuck charged that the Quebec contracts were not only outrageous and inequitable, but also illegal and unenforceable. The latter contentions, which caused chaos in the bond markets, were based on two premises: that the interprovincial character of the contracts placed them in federal jurisdiction and, therefore, beyond the capacity of two provinces to negotiate; and that, as the commission was a municipal cooperative, the contracts lacked the consent of the municipal hydro commissions. While the first justification appeared to be an interesting point of law, the second was fallacious because the Power Commission Act clearly left decisions on supply to the HEPC, with no statutory municipal role.[45]

Although the legal merit of Roebuck's grounds was dubious, this was lost in the controversy sparked by the threat of repudiation. The discord was pronounced because although he had been preparing his case against the contracts since the election, he had done so without revealing what action might be expected. Repudiation was even a surprise to the cabinet, to which he had not disclosed his intentions until, in his own admission, the very last moment. As a result, his addresses left the cabinet divided, with Nixon, among other senior members, believing the contracts were valid. Hepburn had previously told Roebuck that he wanted to settle the dispute peaceably, but the premier was campaigning in northern Ontario when the controversy erupted.[46]

On Hepburn's return to Queen's Park on 28 February 1935, the cabinet met to discuss repudiation. At the close of the meeting Hepburn revealed publicly that the cabinet had yet to take a position on the legality of the contracts, but he added that the opinion of Roebuck would carry great weight. Following another meeting on 4 March, Hepburn declared that 'the Cabinet to a man [was] right behind Mr Roebuck.' The cabinet was not aware, however, that Roebuck had not been forthcoming about the HEPC's power supply. He had failed to disclose that Stone and Webster, a Boston-based engineering consulting firm, had informed him in October 1934 that expected improvements in the economy would create a shortage in the event of repudiation.[47]

Roebuck also made his case against the contracts in a hyperbolic four-part address on CFRB radio in late March. Arguing primarily that the Quebec contracts would have been unnecessary even if the depression had not occurred, he invited listeners during the first three addresses to write to him so that he could be guided to a 'just and courageous decision' on whether to accept the burden of these 'improvident' contracts or take action. In the fourth, he offered to assist the listeners in coming to the conclusion that the contracts were void. At the OMEA's request, Gaby would dispassionately respond to Roebuck's criticisms on CRCT radio on 6 April.[48]

Undoubtedly as a display of government unity, Hepburn himself introduced the bill to declare the Quebec contracts 'illegal, void and unenforceable.' When debate began on 8 April 1935, both the government and opposition engaged in a pitched battle that lasted around the clock until the legislation was passed on 11 April. The intent of the act was clear, but it provided leverage for renegotiating the contracts through two provisions: it barred legal reprisals, and it permitted revised contracts subject to cabinet approval. In addition, the act also assured that power would be supplied if needed by providing for delayed proclamation. This last provision, however, only surfaced by amendment during debate. It resulted after Hogg, the chief engineer, hydraulic and operation, had informed Lyon that the HEPC was ill-prepared to operate without Quebec power. At this point Lyon had to inform Hepburn that Roebuck had kept himself and Hogg in the dark. Hepburn then had Stone and Webster evaluate Hogg's view. In July the company confirmed Hogg's assessment, stating that Ontario would face a power shortage by late 1937 if the contracts were cancelled and no thermal plants were constructed.[49]

In the midst of the high politics surrounding the contracts, McCrimmon, the controller, delivered a confidential report to Lyon on the HEPC's organization. Besides making recommendations for improvements along business lines, the centrepiece recommendation was that the commissioners be relieved by management of all purely administrative matters so that they could devote their attention to general policy. In suggesting such a modern corporate division of labour, McCrimmon felt that the commissioners needed only to meet every fortnight, or even monthly to consider reports from management. Given the direct political control of two ministers on the HEPC and the seriousness of the dispute with the Quebec companies, his recommendation for such decentralization would not be followed. Indeed, Roebuck, in the margin of his otherwise unmarked copy of the report, wrote 'NO' beside the recommendation.[50]

When threats alone did not prove sufficient to entice the Quebec companies to renegotiate their contracts, the HEPC made motions toward repudiation in the fall of 1935. On 19 September two of the companies were notified that the additional power slated for delivery in their contracts on 1 October would not be accepted. The next day Hepburn received an assurance from Stone and Webster that although the commission would be at some operational risk without Quebec power, it could meet the probable consequences. Hogg disagreed with this conclusion and, by memorandum to the commissioners, warned that the HEPC would only be able to operate with interim emergency service under hazardous conditions. Roebuck brought the memorandum to cabinet on 17 October, but with the misleading notation that Hogg believed the commission could function without Quebec power. This advice paved the way for the cabinet to approve repudiation.[51]

On 18 October the HEPC commissioners – Lyon, Roebuck, and McQuesten – voted unanimously to stop taking power under the Quebec contracts as of 22 October. They also recommended to the cabinet that the repudiation provision of the statute be proclaimed in order to cancel the contracts outright. There followed a meeting between the four Quebec companies and the full Ontario cabinet on 23 October. The companies, for their part, asked for and received time to come forward with concessions. Unbeknownst to the others, Gatineau then made an informal agreement for the delivery of power, resumed before the end of October. Thereafter, Beauharnois, MacLaren, and Ottawa Valley offered to write off 45 per cent of the HEPC's losses on each of their contracts. At a second meeting with the cabinet on 4 December, Roebuck rejected the write-off proposal on the grounds that it required the commission to divulge data on consumption. Hepburn then informed the companies that the repudiation provision would be proclaimed that day. Since the companies could not revise the terms of their contracts without the consent of their bondholders, repudiation was needed, he informed them, to force their bondholders' hands.[52]

Although repudiation affected all the contracts equally, Lyon and Roebuck later revealed that the HEPC had decided beforehand to arrange new contracts with all the companies except Beauharnois. Towards this end, tenders were invited from all four companies, but terms were arranged with only Gatineau and MacLaren on 20 December 1935. Gatineau's tender set the pattern, offering to supply power at $12.50 per horsepower instead of $15.00 and to accept liability for increases in Quebec taxation. In exchange, Gatineau asked for and received a ten-year preferential supply arrangement whereby the HEPC would assume the full complement of its previous contract before any other Quebec power would be purchased. Facing insolvency, MacLaren revised its offer to meet Gatineau's price, although for a fixed supply of less than one-third of the power in its previous contract. Its contract was officially signed 1 February 1936, one week before Gatineau's on 8 February, presumably to obviate the latter's supply preference.[53] With the OMEA convention falling in the midst of this battle over repudiation and renegotiation, Lyon broke the annual tradition of the HEPC chairman addressing the delegates.

Since the tenders of Beauharnois and Ottawa Valley were rejected by the government, these companies moved to seek legal redress on the basis that the repudiation Act was *ultra vires*. A two-year legal conflict over the contracts followed. The first sign of victory in the struggle came for Ottawa Valley on 19 November 1936 when the Ontario Court of Appeal found that a provincial legislature was 'powerless to derogate from civil rights outside the province.' Hepburn called the decision a temporary hollow victory and stated that his government, 'in the public interest, will fight to the finish and, I repeat, never pay.' As was expected, Beauharnois would win its suit on 13 January 1937 in the Supreme Court of Ontario, at which point the HEPC would appeal the decision in Ottawa Valley to the Judicial Committee and the decision in Beauharnois to the Ontario Court of Appeal.[54]

In order to respond to the HEPC's court losses, the cabinet decided on 4 January to recall the legislature on 19 January, three weeks earlier than usual. The purpose of the special session was to permit Roebuck to introduce three pieces of legislation that would ensure that the HEPC was beyond the jurisdiction of the courts. His objectives were to reaffirm that the consent of the attorney general was required for suit against the HEPC; to shield commission property from court awarded damages; and to exempt it from having to provide a security deposit for Judicial Committee appeals.[55] In the midst of this flurry of activity, Hepburn left for Arizona to recuperate from bronchitis, leaving Nixon as acting premier.

Given the power shortage that had been predicted for the winter of 1937–8, the government's continuing dispute with the Quebec companies was becoming an embarrassment. Recognizing the political consequences of a power shortage, the government sought to ensure an adequate supply, but Lyon and Roebuck ruled out further Quebec contracts. They were adamant in their opposition because their alternative, the back to Niagara scheme, was now more than rhetoric. It involved redirecting water from the Hudson Bay watershed into Lake Superior with the extra flow at Niagara increasing the HEPC's output by 100,000 horsepower. But despite the promise of inexpensive power, an important impediment nevertheless stood in the way of this plan. The United States, following Roosevelt's re-election in November 1936, had been pressing Canada for a St Lawrence development agreement. Since Ontario was needed as a party to develop the river's power, International Joint Commission approval for the diversions alone would not likely be forthcoming. In the face of these pressures, Roebuck and Hogg had met with federal officials on 14 January 1937, where Roebuck audaciously suggested that the HEPC might proceed with the diversions without such approval.[56]

Unable to wait for approval and construction of the Hudson Bay diversions to secure a reserve margin of power, Hogg directly explained his plight to Nixon on 20 January. Much to Lyon's consternation, Hogg told Nixon that the HEPC should come to terms with Ottawa Valley, which, despite the commissioners' pronouncements to the contrary, was prepared to settle. On 22 January, with Roebuck pressing his bills through the legislature, Nixon sent a note to Hepburn in Arizona, together with a report from Hogg to apprise him of the situation. House leader Harold Kirby also wrote independently that the caucus favoured a settlement. On 24 January Hepburn consented to the negotiations and insisted that the two most controversial of Roebuck's bills, the shield from court awards and the security exemption for Judicial Committee appeals, should be amended to permit delayed proclamation. With these changes, the legislation passed 29 January, at which point the session adjourned.[57]

Despite this turn of events, the HEPC only reached new terms with Ottawa Valley after further political interventions. At Nixon's request, the company's representatives came to Toronto on 1 February, but Lyon refused to meet with them on the grounds that Nixon had pre-arranged the terms of the settlement. In writing, Nixon ordered Lyon to negotiate, and Roebuck and McQuesten, the two ministers on the HEPC, had to be pressured as well. An initial agreement was worked out on 4 February, but the negotiations bogged down, with Lyon insisting on an option to buy the company, and the company refusing to absorb increases in Quebec and in federal taxation. Both relented in the agreement forged on 12 February, but only after Hepburn had cut short his stay in Arizona to return and break the deadlock. In reaching this settlement, the government had to abrogate Gatineau's preferential supply arrangement of 1936. The new Ottawa Valley contract was drafted as a revision to its earlier contracts of 1930 and 1931 and was ratified 'notwithstanding' any other act in order to circumvent Gatineau's supply preference.[58] Ironically, it was during these negotiations that the new OMEA president, F.C. Elliott, lectured his membership on the need to have 'no political bias in dealing with Hydro matters.' He even signalled his desire to improve the OMEA's poor relations with the government, stating that 'only by our interest in Hydro matters can we justify our existence.'[59]

While Roebuck's and Lyon's opposition to purchasing power in Quebec was well known, it did not yet place them on the wrong side of the government's new pragmatism on Quebec contracts. In February 1937 Hepburn committed himself to their Hudson Bay diversion plans after Roebuck and federal officials, at the behest of the Roosevelt administration, had resolved the impasse over Ontario's objections to a St Lawrence treaty. Under the treaty's proposed terms, the HEPC would be permitted to proceed with the diversions immediately and to delay its power development obligations on the St Lawrence for up to ten years. However, in March Hepburn informed Mackenzie King that the treaty would have to wait. He realized that the HEPC would not need the power from either the diversions or the St Lawrence if the impending decision of the Ontario Court of Appeal went in favour of Beauharnois. Also, he feared that the treaty would publicly reveal that Ontario faced a power shortage. For these reasons, he wanted to hold an election before making the policy reversal. The Roosevelt administration agreed to hold the treaty over until 1938.

Although back to Niagara ostensibly remained the government's policy, Roebuck, the leading proponent of this scheme, resigned from cabinet on 14 April 1937. He did so at Hepburn's request. The pretext was his alleged breaking rank on the premier's campaign against American industrial unions. However, Roebuck had also lost credibility with the cabinet and caucus over hydro policy. While Hepburn did not request that Roebuck resign as a HEPC commissioner as well, Roebuck did so later in April, believing that to continue would leave the wrong impression of his relationship with the government. Later, he would claim that he had been forced to resign from both the cabinet and the commission because he had refused to be party to a new contract with Beauharnois.

The government needed to avoid a loss of its Beauharnois appeal, and the HEPC needed the power to avoid a shortage. Both sentiments were expressed directly to Hepburn on 8 June by Jeffery, the chief engineer of municipal relations and rural power. Nevertheless, the contracts were not renegotiated at this time because Hepburn feared that it would lead Roebuck to campaign against the government in the forthcoming election. Presumably to ensure his silence, Roebuck was given a retainer the next day for the HEPC's expected Beauharnois appeal, having first been offered the case in May.[60] After the Ontario Court of Appeal sided with Beauharnois on 22 June, Hepburn moved to minimize the defeat on three fronts: the decision was appealed to the Judicial Committee even though he was advised that the HEPC would lose there as well; Roebuck's legislation shielding the HEPC was proclaimed; and as a public display that government policy had not changed, construction on the Hudson Bay diversions began in July, even though they lacked International Joint Commission approval.

Although the prospect of a power shortage was an issue raised by the Conservatives in the October 1937 election, it did not affect the outcome, as the Liberals were returned with a second majority. Hepburn had deflected the opposition's criticism during the campaign by drawing attention to the financial advantages of repudiating and renegotiating the Quebec contracts and by extolling the virtues of the back to Niagara scheme. Ten days after the election, however, the policy of the government changed when Beauharnois, recently acquired by Herbert Holt's Montreal Light, Heat and Power, served notice on 16 October that it wanted to negotiate a new contract. Hepburn agreed to these negotiations, but Lyon obstinately stood in the way. As a result, Hepburn asked for Lyon's resignation. Although McQuesten, the third and only other appointee, continued as minister of highways, he was also removed from the HEPC, the only apparent reason being a change of policy. McCrimmon, the secretary and controller, also resigned at this time, making way for Hepburn to allow his new commissioners a free hand.

Economic Stability and War Mobilization

In remaking the HEPC through a wholesale change of commissioners on 1 November 1937, Hepburn did not make the appointments with the partisanship that he had shown in 1934 and Henry had practised in 1930. Instead, he implemented a variation on the formula begun by Ferguson of a non-partisan chairman and government and OMEA constituency representation. He appointed Thomas Hogg chairman, thus giving the commission an inside rather than an outside chairman like Magrath and Lyon. Hogg had been offered the position at a cabinet meeting on 26 October, where he responded that he needed the latitude to settle the Beauharnois contract. Hepburn rejected this as a condition, offering instead permission to explore the issue.[61] Hogg also became the sole chief engineer at this time, meaning the division of labour between chief executive officer and chief operating officer(s), which had been rife with conflict for Lyon and Hogg, disappeared. Jeffery, who had been chief engineer of municipal relations and rural power, became chief municipal engineer as well as acting secretary and controller. In 1938 Osborne Mitchell would become secretary, without anyone officially designated controller thereafter.

To fill the remaining commissioner posts, Hepburn appointed William Houck, a minister without portfolio since the election and the Niagara Falls MPP, as the statutory cabinet representative. In making the other appointment, he judiciously brought the OMEA back into the HEPC's decision-making. Having repeatedly been called upon by the OMEA to renew its representation on the commission, Hepburn asked the association for a nominee, as Ferguson had done in 1925. Although the nominee was not chosen, the OMEA later approved of the appointment of Albert Smith, the Liberal MPP for Waterloo North. Smith was a vice-president of the OMEA for 1936–7 and also the past president of the Ontario Association of Mayors.[62] By placing one cabinet minister and one backbencher on the HEPC, Hepburn had maintained, although in somewhat diminished fashion, his capacity to influence a majority of the commissioners.

Although both Lyon and Roebuck were out of the way, their departure had not brought an end to Hepburn's troubles. Hogg informed Hepburn on 18 November 1937 that a settlement with Beauharnois would leave the HEPC with a new power surplus. Moreover, to effect a settlement, Hogg had to offer Gatineau a new long-term contract for it to relinquish its preferred supplier status, and in fairness he granted similar terms to MacLaren. To absorb the surplus, Hogg made arrangements to sell power to private companies in the United States, and Hepburn, for his part, asked the federal cabinet for an export permit on 29 November. King, however, refused to take this direct responsibility for the politically sensitive matter of exporting Ontario power. After Hepburn complained publicly, King chose to deflect responsibility by having the Electricity and Fluid Exportation Act amended to require export permits to proceed by way of a private member's bill. With the Beauharnois and other Quebec contracts having been finalized on 14 December, Hepburn formally applied for a permit under the new rules on 21 January 1938, but this application would run into a roadblock as well.[63]

In a setting where Lyon and Roebuck had publicly criticized the new Quebec contracts, Hogg addressed the HEPC's abrupt change in policy at the February 1938 OMEA convention. Contrary to received wisdom, he informed the delegates that growth in demand was expected to outpace supply in the short term. His calculation for the 'probable best provision' for increasing supply, the point where the consequences of a shortage outweighed the cost of a surplus, was 7 per cent. This was close to the 7.7 figure used in 1929 before signing the original Quebec contracts. Under these circumstances, Hogg explained that the HEPC had only two options – either steam power, with its requisite disadvantages, or Quebec power. He emphasized that the latter was selected strictly on the basis of demand and cost, but he added that it had an important advantage. The resolution of the Quebec contract disputes was a necessary first step in facilitating the interprovincial cooperation that would be required to develop the Ottawa River. In closing, Hogg stated that the new contracts were consistent with reports he had forwarded and seen rejected by the former commissioners. For this reason, he argued that they 'alone must accept full responsibility for the past policies.'[64]

In addition to negotiating a new contract with Beauharnois, Hogg had also repaired the HEPC's relationship with the municipal hydro commissions. This was much in evidence at the 1938 OMEA convention. Houck made a conciliatory speech, stating that there was 'only one hydro scheme' with its provincial and municipal sections differentiated by 'purely mechanical matters of jurisdiction and administration.' In his view, this meant that cooperation in the performance of their separate duties was essential. President Elliott, for the OMEA's part, was able to state that his executive had already received a cordial reception from the HEPC commissioners and added that the appointment of Smith as the third commissioner would promote harmony. President-elect Gordon Matthews also indicated a measure of the HEPC's goodwill when he announced that the association, which had never had an office, had received permission to hold its executive meetings in the HEPC's boardroom. Given this turnaround in relations, the OMEA pledged its cooperation with the new commissioners and expressed its confidence in Hogg. Likely as a gesture of good faith, Matthews also gave notice of motion to limit the tenure of the OMEA president to two one-year terms. The intention was presumably to guard against long-lived personal leadership such as that during Maguire's twelve-year presidency, a situation which had irritated the Liberals while in opposition. This resolution was later adopted at the 1939 convention.[65]

As for the fate of Hepburn's application for an export permit, King announced on 21 March 1938 that the bill had been rendered moot. The Roosevelt administration would not permit the importation of the power, fearing that it would undermine plans for public power developments in New York State and leave international power-pooling in private hands. The United States, however, was willing to accept the power under a comprehensive St Lawrence treaty. Under these circumstances, Hepburn rejected the offer and accused King of participating in a conspiracy with Roosevelt.

During the 1938 legislative session Hepburn also faced trouble from another quarter. Conservative leader Leopold Macaulay had charged that

Hepburn had sanctioned surreptitious negotiations with Beauharnois prior to the 1937 election and had deliberately deceived the electorate with the back to Niagara scheme as a cover. A select committee was appointed to investigate the charges on 28 March. After hearings that summer, the committee's report in the spring of 1939 was divided on partisan lines. The majority report exonerated Hepburn on two grounds: that Beauharnois officials had testified that secret negotiations had not taken place; and that Hogg had testified that Lyon knew of the need either to renegotiate the contract or develop steam power but had not informed Hepburn of Hogg's reports. The minority report concluded that Hepburn either knew or should have known of the power shortage.[66]

With war on the horizon, the renegotiation of the Quebec contracts and the renewal of good relations between the OMEA and the HEPC became an essential element of industrial mobilization. Supplying power for the war effort was the commission's new preoccupation, as was the concomitant need for additional sources. This requirement became evident through three developments in the summer and fall of 1939. First, Hogg's speech to the OMEA's summer convention emphasized the 'mutual interdependence' of the HEPC and the municipal hydro commissions to ensure their cooperation for war preparedness. Secondly, the Power Control Act was passed in a rare fall session of the legislature on 22 September. It gave the HEPC emergency powers to regulate and control all aspects of the electric industry, especially the allocation of power to end users.[67] Thirdly, the St Lawrence seaway and its attendant power developments received Hepburn's endorsement on 3 October, although his timing was wrong for a treaty with Roosevelt entering an election year. Nevertheless, the United States did remove its objections to the HEPC making use of the extra water from the now completed Hudson Bay diversions, leading to International Joint Commission approval in September 1940.

Following Roosevelt's 1940 re-election, an international agreement on the St Lawrence was reached in January 1941 and a federal-provincial agreement in March. However, opposition from powerful economic and sectional interests in Canada and the United States ensured that both the seaway and the power development projects would not go ahead until the 1950s. After hope for the ratification of the treaty had passed in 1942, Hepburn authorized Hogg to negotiate the development of the Ottawa River with the governments of Quebec and Canada, and the agreement was finalized in January 1943. Through 999-year leases by one province to another, it allocated the full benefit of the lower river's power sites, including the previously contested Carillon Falls, to Quebec and the upper river's sites to Ontario.[68]

The Ottawa River agreement was ratified in the legislature on 19 February, but not before George Drew, the new Conservative leader, had accused the government of selling out Ontario's interests. In particular, Drew accused the government of bartering away Ontario's half of the Carillon site for, in his view, sites of dubious value. He also chastised the government for receiving no assurance in return that Carillon would not be developed by private interests for the export market. In addition, he criticized the long-term leases on the grounds that they were grants, cautioning that no province had the constitutional power to make such transfers and that no future government could be bound by such terms.[69]

Although Hepburn had initiated the intergovernmental negotiations for the development of the Ottawa River, he was no longer premier when the agreement was finalized. He had stepped down unexpectedly on 21 October 1942, personally designating Gordon Conant his successor after Nixon refused to accept the leadership of the party without a convention and Farquhar Oliver refused without a vote of caucus. Under pressure from caucus, Conant later agreed to a leadership convention, which Nixon won on 30 April 1943. With an election supposed to have been held by October 1942, but postponed for one year by an all-party agreement, Nixon called one for August 1943. In the results, the Liberals received only fourteen seats to the Conservatives' thirty-eight and the CCF's thirty-four. Thus began the Conservatives' dominance of postwar Ontario politics.

5

Infusing the Hydro-Electric Power Commission with Policy Sensitivity: One-Party Dominance, 1943–63

From the 1943 election to the resignation of the last minister to sit as a commissioner in 1963, the Hydro-Electric Power Commission emerged as a full-blown facilitative crown corporation in a booming postwar economy. In this economic environment the Ferguson formula for appointments – a non-partisan chairman alongside government and OMEA constituency representatives – hindered the government's efforts to infuse the HEPC with policy sensitivity for its economic objectives. The result was a protracted tug-of-war over the institutionalized ambivalence of the commission as both a government corporation and the trustee of a municipal cooperative. Given a political environment of one-party dominance, the government resorted to statutes, its instrument other than appointments for setting the direction of the HEPC, as a tool in its struggle. The OMEA, for its part, resisted the government's ambitions through convention resolutions, which carried political weight because the association was broadly representative of the province.

Although the power to appoint had been used extensively by government to manage the HEPC's affairs, its utility in the face of new pressures caused by postwar economic growth was constrained by the Ferguson formula. The three major projects that this growth occasioned – frequency standardization, redevelopment of Niagara, and development of the St Lawrence – and the development of the Ottawa River already under way had an unprecedented economic impact. Their combined effect was to place the commission beyond the scope of the municipal cooperative myth, despite the OMEA's continued contention to the contrary. Thus, the government made extensive use of statutory enactments to authorize HEPC industrial leadership in an effort to weaken the association's influence upon the commission's affairs.

The government attempted first to weaken the Ferguson formula by creating an advisory council for the HEPC, but this effort failed because of OMEA resistance. Then the government increased the number of commissioners but at the price of conceding increased and disproportionate constituency representation to the association. Subsequently, the government, circumventing the OMEA, created an energy portfolio, but it was not made responsible for the commission because of association objections. After achieving only limited success in exposing the HEPC to outside input, the government, in exasperation, increased its own presence on the commission and bolstered it by appointing a senior civil servant. The government then removed the statutory requirement for ministerial representation, otherwise leaving the HEPC's decision-making structure unreformed. When the last minister resigned in 1963, the government's struggle with the OMEA was essentially at a draw.

Planning for Postwar Economic Growth

Nine years after being ousted by the Liberals under Hepburn, the recently renamed Progressive Conservatives returned to power in 1943 with a minority government under George Drew. In the electoral campaign, they had made the 'removal of the HEPC from the political arena' one of the twenty-two points of their platform for postwar reconstruction.[1] This campaign promise was kept, in part, when Drew reviewed the appointments made by the Liberals to the HEPC. On 24 August he removed cabinet representative William Houck and OMEA representative and Liberal backbencher Albert Smith, both of whom had been defeated in the election. However, he did not remove Thomas Hogg, who, while continuing as chief engineer, had been appointed chairman by Hepburn.

Drew was likely not in a position to remove Hogg, given that he had only a minority government in a situation where the HEPC was vital to the war effort. He nonetheless had two other appointments through which to exercise influence over the respected chairman. In late August, George Challies, who like Drew had publicly criticized Hogg over the Ottawa River agreement, was appointed one of the commissioners. As a minister without portfolio, he filled the statutory requirement for one to two cabinet ministers. A third commissioner was not immediately appointed. By not selecting an OMEA representative, Drew was attempting to break away from the Ferguson formula for appointments. He would not succeed, however, because the formula had reached the status of a semi-convention with Hepburn's return to it in 1937.

The exclusion of the OMEA in Drew's appointments to the HEPC became an issue at the association's February 1944 convention. A report and resolution from an ad hoc Committee on Municipal Representation called on Drew either to give the association representation or to grant it a role in shaping HEPC policy. Drew, in an unprecedented speech to the convention, chose the latter course. He announced that an advisory council, on which the municipal hydros would be 'well and adequately' represented, would be created to assist the HEPC commissioners plan for the postwar future. The announcement was heralded as 'one of the great forward steps' in promoting cooperation between the commission and the OMEA and was accepted as having met the spirit of the representation resolution. Fulfilling the promise, Drew had the Power Commission Act amended on 14 March to create the Ontario Hydro-Electric Advisory Council as a body of five members appointed for two-year terms by order-in-council. Although its creation did not alter the powers of the HEPC commissioners, the council was mandated to make reports for their 'consideration and assistance' on subjects it chose or was assigned. However, having formulated this limited mandate, Drew did not proceed to make appointments to the council.[2]

The mystery of Drew's non-appointments to the advisory council was answered in part when Ross Strike, the newly-elected OMEA president, was appointed the third HEPC commissioner on 16 June. Strike's son Alan recollects that his father had been calling for association representation and had likely lobbied Leslie Frost, the provincial treasurer, and possibly Drew directly, although not on his own behalf. Strike could do so because he had important connections with the new Conservative government. He had been a personal friend of Frost's since their days in a special Osgoode Hall Law School class for war veterans in 1921, and he had been a Conservative candidate in the 1941 federal election.[3]

In appointing Strike, Drew had bent to all of the Ferguson formula's rules, although the variation remained that the chairman came from inside rather than outside the HEPC. With his government returned with a majority in June 1945, the status of the formula as a semi-convention was secure. Regarding his own role in this arrangement, Strike stated in his farewell presidential address to the OMEA on 5 March 1946:

> I was instructed by the Premier of this province that I was to consider myself as the representative of the municipalities on the Provincial Commission with no strings attached as far as the Government was concerned. I immediately dropped all my political affiliations so that as far as possible the municipalities would be independently represented. This position has also been recognized by the Provincial Com-

mission and I am treated as the direct representative of the Municipalities in all matters that may affect them.

Strike nevertheless recognized that his dual responsibility as OMEA president and HEPC commissioner, a situation which had not existed since Maguire had held both positions from 1925 to 1934, presented him with conflicting loyalties. He stated to the delegates that for the OMEA to be 'completely free and unfettered' and without an appearance of being dominated, its president should not, in future, be a HEPC commissioner.[4]

In his speech to the March 1946 OMEA convention Hogg outlined the HEPC's two postwar challenges, the most pressing of which was the need to expand the power supply. Making an admission of miscalculation, he stated that there had been a continuous increase in demand while the HEPC had hoped for a recession load. The other challenge was the need to proceed with the standardization of electrical frequency in Ontario at 60 cycles, the new North American norm, by converting the 25-cycle portions of the province. Although the government supported both these initiatives, there was now a discernable indication of displeasure with Hogg. Following the convention, the Power Commission Act was amended on 27 March to provide for one of the commissioners to be a vice-chairman with all the powers of the chairman if there were an absence, illness, or vacancy.[5] Challies was made vice-chairman, and within a year Hogg would be fired over the power shortage.

Hogg's ouster resulted from his irresponsibility in handling the power shortage during the winter peak demand of 1946–7. For Drew, it was the last straw in Hogg's repeated failure to 'communicate essential information' to the government sufficiently in advance of when the government would be required to take executive or legislative action. The incident began on 3 December 1946 when the HEPC gave public notice of the need for voluntary conservation of electricity to stave off a shortage when the winter peak arrived. The intimation was that conservation was preferable to emergency rationing under the wartime Power Control Act of 1939. When the voluntary measures failed, the commission informed the government on 19 December that the emergency powers were necessary. Knowing no more than the public, a perturbed Drew later stated that he had not been forewarned of such a serious problem, although he did not explain why Challies, the cabinet representative on the HEPC, had not provided the information as the official liaison. Drew was also upset that the notification from the commission was not accompanied with specific recommendations, leaving only the impression that they would be forthcoming. Such specifics were, he added, an essential requirement under the act.[6]

In a situation where press reports were warning of imminent emergency rationing, Drew wrote to Hogg on 3 January 1947 asking for advance knowledge of the HEPC's forthcoming recommendations. He also wished to be informed of whether any power could be obtained from Quebec or American sources. Presumably unbeknownst to Drew, Hogg had left for four weeks vacation in the Bahamas on 21 December, two days after the announcement of the emergency. To make matters worse, Hogg was known to be travelling with a number of prominent Liberals, including the federal minister of reconstruction, C.D. Howe, and Senator Norman Lambert, both key party strategists. This was at a time when Drew was engaged in a ongoing conflict with the federal government over federal-provincial finance. In Hogg's absence, Challies and Strike, the other two commissioners, received Drew's letter. To their surprise, they discovered that directors of a subsidiary of Niagara-Hudson, an American company, had offered 100,000 horsepower to the HEPC directly through Hogg on the day before he announced the emergency. The two commissioners then made immediate arrangements to purchase the power and rescind the request for emergency powers.[7]

On Hogg's return to Toronto, Drew privately requested the chairman's resignation. The terms were extremely favourable according to Drew, and given that Hogg was sixty-three years old, the departure would appear publicly as an early retirement. However, while the arrangements were being finalized, the press got wind of the story, and Hogg did not deny its authenticity. On 24 January the *Evening Telegram* and the *Globe and Mail* both devoted a few column inches to Hogg's comments that the matter was 'entirely in the hands of the Prime Minister' and that 'any announcement ... would have to come from the government.' For its part, the *Toronto Daily Star*, in a banner front-page headline, attributed Hogg's looming dismissal to two policy disputes with the premier, both of which Drew later called malicious fabrications. The first was Drew's presumed objections to the 1943 Ottawa River agreement Hogg had negotiated with the Quebec and federal governments. The second was frequency standardization, a scheme which Drew was said to favour and Hogg was incorrectly said to oppose. The *Star* added in another front-page story the next day that the power shortage would not have resulted if Drew had not interfered with Hogg's plans for development of the river. Worse still, the firing of Hogg was said to presage 'political control' by Drew or the creation of a department of power in government, both historically a call to arms for the OMEA.[8]

Seeking to diffuse a major political controversy, Drew took action on 27 January 1947 through a province-wide address on CBC radio. He began by explaining that although the close relations of the provincial and municipal hydro systems left the impression that they were one system (cooperatively owned by the municipalities), the province and each municipality were separately responsible for the management of their respective systems. Moreover, he stated that with the whole social and economic life of the province dependent on the HEPC, the government had no higher responsibility than to ensure that the provincial commission provided sufficient power resources for continued industrial development. As an illustration that the poor communication displayed in the power shortage was not an isolated event, Drew related the embarrassment the government had felt on another matter in the fall of 1946. The HEPC's proposal to purchase Ottawa's private power company was debated by Ottawa city council after Drew had been assured by Hogg that the negotiations had not proceeded beyond the discussion stage. As a result of Hogg's actions, Drew stated he had no choice but to remove him in the public interest. In order to dispel concern over the *Star's* warning of political control, Drew announced that he intended to strengthen and enlarge the HEPC in the 1947 legislative session, a statement likely added to appease the OMEA by holding out the prospect of increased representation.[9]

In the days following Drew's dismissal of Hogg, newspaper editorials took up the subject, choosing sides on the old question of whether the HEPC was a government corporation or the trustee of a municipal cooperative. The *Telegram*, defending Drew staunchly, argued that the government was directly responsible to the people of Ontario as the 'owners' of the HEPC. It stated that the government had a right and a duty to remove commissioners when necessary and more generally to guard against power shortages in the future. Its only criticism was that Drew waited too long to make his radio broadcast, allowing the 'rag-tag and bobtail' of his opponents time to 'sing their chorus of hate.' In a more moderate defence of Drew, the *Globe* described his actions as clearly justified given that the hostility between the government and the HEPC could not be allowed to continue where 'sincere and ready cooperation should be the rule.'[10]

Taking an opposing view, the *Star*, over three days, criticized Drew on the grounds that the HEPC commissioners, although appointed by the government, 'act as trustees for the Hydro municipalities which own the system.' Without ever mentioning the power shortage, the paper dissected Drew's example of the negotiations for the Ottawa company and lauded the engineering accomplishments of Hogg. On the former, it contended that hydro was entrusted to autonomous commissions to preclude political interference, stating that the provincial and municipal hydro commissions would be paralysed if they had to consult the government on every important decision. In addition, the *Star* wondered why Challies, the minister without portfolio on the HEPC, had not kept Drew informed. The paper, which considered Hogg to be held in universal esteem, argued that he '[did] not need to be told by politicians what [was] good for the Hydro public and the Hydro municipalities.' And to stir the pot, it stated that Hogg's dismissal for not dancing to Drew's tune should not go without protest.[11]

Following the barrage of editorials, the release of Hogg's letter of resignation and Drew's letter of its acceptance helped to clarify what had transpired. While Hogg conceded that he could respect and understand the government's desire for a change in the direction of the HEPC, he added that his removal was a relief because his general health was not what he desired. With mild regret, Drew responded that the 'utterly unprincipled conduct' of the *Star* in speculating on the reasons for the dismissal had left him no choice but to go public with the problems between them.[12] Despite these protestations, the *Star* was correct on one of its speculations. On 30 January, the day before writing back to Hogg, Drew had made an agreement with the Quebec premier, Maurice Duplessis, altering the terms of the 1943 agreement that Hogg had negotiated for the development of the Ottawa River. In essence, the revised agreement kept the previous allocation of power sites, but it respected Drew's objections to one province being granted property in another.[13]

At the OMEA convention in March 1947 what remained of the controversy over Hogg's removal was dissipated by high-level conciliation. President R.M. Durnford acknowledged that the government had the prerogative to 'impinge' on the powers of the HEPC commissioners, but he urged his membership not to seek changes to the Power Commission Act based on partisan considerations. In his view, this would 'get us exactly nowhere, except to lower the prestige, dignity and influence of our association.' Drew, in his second speech to the association in three years, repeated the main points of his radio address but added that he did not intend to depart from the municipal cooperative principle. Strike then urged the delegates not to be guided by 'sloppy thinking.' He informed them that the estimates of postwar demand in Canada and other countries, based as they were on experience after the First World War, had been wrong. The controversy came to an end when no action was accepted on motions calling for Hogg to respond to Drew's charges and for Challies to explain why he had not informed Drew of the shortage.[14]

As the controversy over the firing of Hogg was dissipating at the convention, a new one was brewing over the HEPC's interim plans for frequency standardization. Standardization, which had been contemplated before the war, was needed to link the commission's generating stations into a fully integrated transmission network for postwar development. This situation arose because the HEPC and the private companies it had purchased had begun their operation with different frequencies. The general economic efficiency of integration and the growth in use of electrical appliances and machinery, which were not transferrable from frequency to frequency, became the justification for standardization. The dilemma for the OMEA was that of the estimated cost of $195 million for the fifteen-year project, the affected municipal hydro commissions would be required to pay $35 million for the conversion of their own equipment. Moreover, the HEPC, as Durnford pointed out, could make the final decision to proceed whatever the views of the municipal hydros. Until the day of decision arrived, Durnford announced that the OMEA's immediate objective was to 'be ready for the day when our opinions and cooperation are requested, or extended by our own volition, losing no opportunity in the meantime to advance our knowledge and gain further information.'[15]

The OMEA's lack of knowledge was made all the more apparent when the delegates were informed that external reviews of the HEPC were under way. The commission had hired management consultants Woods and Gordon, financial consultant Geoffrey Clarkson, and engineering consultants Stone and Webster to examine all aspects of its operation, with the financial and engineering consultants paying special attention to the frequency standardization program. The OMEA responded by organizing a special debate on standardization, with J. Clark Keith and E.V. Buchanan, the general managers of the Windsor and London hydro commissions, arguing the advantages and disadvantages respectively. The convention's discussion of the subject ended with a resolution calling for the HEPC to present a final report as soon as possible.[16]

Updating the HEPC's organization, the mandate assigned to Woods and Gordon, was of great urgency. Given that Hogg had served in the capacity of both commission chairman and chief engineer, he had straddled a division of labour that had always existed but had now become more important. Walter Gordon, who later became Canada's minister of finance, in his report outlined a new working relationship for the commission and the government based on this division of labour. His 24 March 1947 recommendations were premised on two interrelated and pressing realities: the retirement of many long-serving employees with intimate knowledge of the HEPC's operation, and the demands of anticipated postwar growth.[17]

To meet these two challenges, Gordon recommended, as had McCrimmon in 1935 while HEPC controller, that the chairman and commissioners be 'concerned with and responsible for questions of policy.' They were to meet twice a month or as required and leave a permanent official, to be styled general manager and chief engineer, responsible to the commissioners for 'all phases of Hydro's operation.' Under this arrangement, and with Hogg's dismissal still lingering in the public memory, Gordon felt that

> The Chairman of the Commission should be directly responsible for liaison with the Government and for keeping the Prime Minister *constantly informed* of the problems, plans and activities of the Hydro. He should see that proper liaison is maintained with the Municipalities, and be responsible for the Commission's 'public relations' in the broad sense. He would be the official head of the Commission, his position corresponding in a way with that of the Chairman of the Board of a large industrial concern.[18]

In essence, Gordon advocated a policy deliberation function for the commissioners, making them more clearly a non-managerial buffer between the government and the HEPC's management. Drew accepted this view by keeping his commitment to expand the size of the commission, but did not follow through with his promise to strengthen the commissioners' powers. The Power Commission Act was amended on 3 April to increase their number to nine from three and dropped the requirement for one to two ministers to sit as commissioners.[19] However, this statutory amendment was never proclaimed, having been left in abeyance after Drew returned to the Ferguson formula. He undoubtedly did so because the OMEA was protecting its municipal cooperative claim, believing appointments made from outside the municipalities were a threat.

Gordon's recommendations for two divisions of labour – the first between the premier and the chairman as chief executive officer and the second between the chairman and the general manager as chief operating officer – were followed only to the extent that the offices were separated after Hogg was fired. One practical reason for doing so was that the dual responsibility required that the chairman be an engineer. When replacing Hogg, Drew did not diverge from the convention of appointing a nonpartisan chairman, although he did leave Challies as the acting chairman while he sought a replacement. In the interim, Challies and Strike, the two remaining commissioners, elevated Richard Hearn from chief engineer, design and construction to general manager and chief engineer in the manner of a chief operating officer. Then, on 1 March 1948, Drew appointed Robert Saunders, a lawyer by profession, as chairman, in the manner of a chief executive officer. In Denison's view, it was Saunders's 'great personal force and imagination' that made him the most attractive candidate. Moreover, as mayor of Toronto from 1945 to 1948 he had publicly advocated frequency standardization, a policy which Drew was actively encouraging.[20] By appointing Saunders from outside the HEPC and alongside government and municipal constituency representatives, Drew had implemented the Ferguson formula for appointments in its original form.

As for Gordon's recommendation to separate policy and administration, no clear divisions developed between the commissioners and the general manager. Saunders, who had no other employment, devoted his full attention to his HEPC responsibilities. Strike, for his part, worked three days a week in the 1940s and four in the early 1950s. Challies worked less regularly than Saunders and Strike, likely because of his responsibilities in the legislature. Moreover, according to Douglas Gordon, then in the HEPC's municipal department and later its general manager, the commissioners met in session once a week rather than twice a month because of the sheer volume of business to be handled in this period of rapid growth.

When the March 1948 OMEA convention opened, frequency standardization dominated the agenda. The HEPC's final report, published for distribution at the convention, recommended that the program proceed. In answer to the municipalities' chief concern, the report argued that the costs of converting their works would be self-liquidating over twenty years as a result of direct savings. Although Saunders, one day on the job, addressed the delegates, it was Strike who explained the HEPC's expectations of the OMEA. He assured the delegates that no attempt would be made to get their consent for the project during the convention. However, with the cabinet having already expressed its desire to pass the necessary legislation, he added that the commission would expect the OMEA to come to a decision by 15 April. This was one day before the legislative session was set to end.[21]

After numerous information sessions and significant discussion, the delegates passed two resolutions. The first established a joint committee of the OMEA and the Association of Municipal Electric Utilities (AMEU), the body of general managers and engineers, to examine the standardization proposal. The second called for the committee's report to be discussed at a special general meeting of the OMEA and for no action be taken by the HEPC or the government before that time. According to Gordon

McHenry, whose father Morris had been in constant contact with the OMEA as director of consumer service for the HEPC since 1938, the association was not being obstructionist. In his view, it was an unwieldy organization that took time to reach a consensus.[22]

On the same day that the OMEA resolution was passed, the speech from the throne announced the government's intention to introduce legislation on frequency standardization. The legislation was subsequently passed by a unanimous recorded vote on 31 March. The next day, the OMEA-AMEU joint committee reported that the standardization program was unnecessary and too expensive. The OMEA executive debated the findings on 5 April, and Saunders and Strike, in answer to its questions, stated that the legislation which had been passed was flexible on the 'mechanics' of implementation. Without taking a position, the executive set the special general meeting for 12 May, unwilling to meet the date Strike had stated the HEPC needed the decision.[23]

The OMEA's timetable had launched it into a showdown with the government. Although only three years into his government's current mandate, Drew called an election on 16 April 1948, the last day of the legislative session. The reason he gave for the early call was that he needed popular sanction for frequency standardization, an action which he declared was part of a 'vast integrated program' of industrial expansion. Although it had received all-party support in the legislature, Drew claimed that the Liberals and the CCF were really against the program. He added that Liberals on the OMEA-AMEU committee studying standardization had orchestrated the negative report to hurt the government. It was rumoured, however, that Drew wanted an election victory to launch his bid to replace John Bracken as national leader of the Progressive Conservative party.[24]

OMEA president George Hutcheson was dumbfounded by both the election call and the use of the HEPC's affairs as the pretext for the election. Since prominent OMEA members were running for all parties, Hutcheson cancelled plans for the special general meeting after consulting members of his executive. Even though he felt he would be censured for the decision, he believed it was advisable for the association to stay removed from party politics. When the full executive met on 3 May, his action was sustained after vigorous debate, with the convention rescheduled for after the election. This decision followed an impassioned speech by Saunders, who exercised his prerogative as honorary president of the association to attend the meeting. Subsequently, the Conservatives received a new majority in the 7 June election, although Drew was personally defeated by a CCF candidate. When the special general meeting convened on 21 June, the

OMEA delegates vented their anger upon the government for proceeding with standardization before the association had made its decision. Although Strike's dual loyalties likely kept him from taking sides publicly at this meeting, Saunders did give a rousing speech on the benefits of frequency standardization. In the end, a resolution, engineered by Hutcheson and sponsored by the executive, isolated and undercut the hard-line opponents. It endorsed the program but urged the HEPC to heed OMEA recommendations on financing.[25]

Struggling over Broader Input in Decision-Making

Although there was speculation that Drew wanted Saunders to be his successor after he stepped down to replace Bracken in October 1948, Leslie Frost won the Conservative leadership, becoming premier on 4 May 1949. Meanwhile, on 1 April two small but significant changes had occurred in the relationship between the HEPC and the government. The Power Commission Act was amended to require that the commission's annual reports be submitted to the provincial secretary, who was then to submit them to the cabinet and formally table them in the legislature. Despite being part of a general scheme to systematize reporting, this change was significant because since 1916 the HEPC had submitted its report only to the cabinet 'for the information of the Assembly.' The frequency standardization provisions of the act were also amended to provide that any new or replaced works were explicitly declared to be the property of the HEPC and were not accompanied by more generous financing provisions for the municipal hydro commissions. This meant that the works could not rightfully be included in any municipal claim to own the HEPC cooperatively. This diminution of the municipal claim continued during Frost's first legislative session. The act was amended on 24 March 1950 to remove the requirement that the annual reports document in detail the sinking fund contributions of each municipal hydro. The HEPC, however, would continue to supply the information and list it as 'equity' in the provincial commission.[26]

Frost had a very definite idea of how the commission's relationship with the government should be structured. He felt that its chairman should be the government's 'right hand in relation to Hydro matters,' even though he recognized that the HEPC had autonomy from the government and that the chairman should not act as a government representative. His reason was that the government carried an 'unavoidable responsibility' for the commission because it 'provided the life blood of the province.' There appeared to be no limit to Frost's interest in the HEPC's affairs; his interventions extended to dictating its statutory appointments to municipal hydro commissions and even to approving the dates of Saunders's vacations. Saunders, however, was not a willing subordinate. In November 1950 he challenged Frost's view, making the case that the chairman should hold the office at 'good behaviour,' like a judge, rather than be 'subjected to the whims or patronage of any political party.' He also felt that the chairman's salary should be commensurate with a full-time position rather than rest at a level which assumed the chairman had other income and business interests.[27] The salary for all the commissioners, $45,000 in total, had not changed since 1924. Nevertheless, Robert Macaulay, whose father Leopold had been Conservative leader prior to Drew and who himself was elected to the legislature in 1951, believes that Saunders was an empire builder like all HEPC chairmen. In any event, Frost chose to make no modifications to the chairman's role while Saunders filled the post.

Frost also faced pressure from the OMEA on the composition of the HEPC. At its February 1951 convention, the first since he had become premier, the association demanded direct representation on the commission. This demand had not been heard since before Ferguson's appointments in 1925. Although thankful that Strike as a former OMEA president was a commissioner, the association's executive was requested to press Frost for an amendment to the Power Commission Act to permit the OMEA to elect one of three or two of five commissioners. Frost did not act on the request and indeed did not even expand the number of commissioners, even though the latter would only have required proclamation of the statute Drew had initiated in 1947.[28]

Following the OMEA convention, an important instance of the government's 'unavoidable' responsibility for the commission's affairs arose. As a result of a Canada–United States treaty and a federal-provincial agreement, a new Niagara Development Act was passed on 5 April 1951. The act empowered the HEPC to construct a second large generating station to meet the increased power demands of the province. The treaty, inventively, removed the maximum restrictions on the amount of water that could be diverted from Niagara Falls, replacing them with minima that had to be maintained during the day for scenic beauty. This enabled the HEPC to divert large amounts of water at night for storage and use the next day.[29]

Given the government's increased responsibility for the commission during the postwar economic growth, Frost rejected the requests of Saunders and the OMEA to relinquish the cabinet's appointment power. Instead, he moved to break the HEPC's concentration of decision-making power by resurrecting the Hydro-Electric Advisory Council, which had been dormant since being authorized in 1944. The Power Commission Act was amended on 5 April 1951 to increase the council's size from five to nine persons. Frost believed that 'the magnitude of Hydro's operations and its influence on the daily life of Ontario's people [made] it imperative that the HEPC should have the benefit of all viewpoints.' For this reason his broad objective in re-creating the advisory council was to counter the popularly held view that the HEPC was beyond democratic control.[30]

Frost was moved to infuse outside advice into the HEPC's decision-making after reviewing private sector corporate organization. Having observed a distinction between a board of directors' responsibility for broad policy questions and its executive committee's responsibility for management, Frost intended to replicate these roles in the HEPC's decision-making. He envisaged that the advisory council, by meeting once a month with the commissioners, would fill a board's function, with the commissioners' daily management function making them akin to an executive committee. Such a diffusion of power, in Frost's view, would improve the HEPC's 'accountability to the public and of public participation in broad policy decisions.' Saunders, at least publicly, lauded the idea as being of great value, but Strike thought the council would be like 'tits on a bull.' According to his son, Strike likely told Frost there was nothing for it to deliberate but that he would not have stood in the way if Frost believed in it strongly.[31]

On announcing his appointments to the advisory council in the spring of 1951, Frost stated that they were chosen for their familiarity with the problems of their respective fields of activity. They were D.P. Bud Cliff, president, OMEA; M.W. Rogers, president, AMEU; C.H. Moors, chairman, Fort William Hydro-Electric Commission; J. Clark Keith, general manager, Windsor Utilities Commission (who had advocated frequency standardization); Marjorie Hamilton, mayor of Barrie (and ex officio member of her hydro commission); J.P. Maher, Toronto Board of Trade; A.F. MacArthur, president, Ontario Federation of Labour; J.C. Brodrick, president, Ontario Federation of Agriculture; and Grattan O'Leary, associate editor of the *Ottawa Journal*.[32] The high profiles of these appointees indicated the importance Frost attached to the council as well as the power of the municipal hydro commissions. In this light, it was surprising that the council never became a presence and indeed may never even have held formal meetings. Macaulay, who was actively engaged in hydro issues as a backbencher and a cabinet minister from 1951 to 1963 and knew many of the appointees, does not recollect ever hearing of the council. He suggests that unless it was a 'private little group of people who met the premier in his bathroom, it had no public status.'

The existence of an advisory council, even with five of the nine members having direct municipal hydro involvement, was likely perceived by the OMEA to be a challenge to its then pre-eminent influence in the HEPC's affairs, if not to its municipal cooperative claim. When the association's executive convened to discuss the council's ramifications, it resolved to request that the government hold the council in abeyance pending an OMEA meeting with Frost. This left Cliff, as association president and a member of the council, in a difficult position because, by accepting the appointment, he had not personally objected to its existence. Cliff later stated that the government had agreed to discuss the principles underlying the council.[33]

Why did Frost not insist that the advisory council function? Denison, acknowledging its inactivity, suggests that the council was created simply to provide a cross-section of public opinion, particularly for the implementation of frequency standardization.[34] For this reason, its continuance might not have been critical. Macaulay, on the other hand, feels that Frost was too thoughtful and farseeing to engage in such public relations gestures. He suggests that the HEPC faced major challenges for which Frost likely thought the council could be of assistance, but two obstacles possibly stood in its way. Saunders, like his predecessors, did not want outside advice because the 'responsibility, affiliation and affinity' of the advisers was to the premier's office, and the advice could be given without his 'knowledge or participation.' Furthermore, the OMEA was a powerful political reality whose opposition could not be taken lightly. With an election looming, a showdown with the association over the council was likely not considered advisable.

When Frost called a provincial election for November 1951 he highlighted his concern over the political implications of the HEPC's affairs for the government. The development of the St Lawrence, historically a politically sensitive project that could have been vetted by the advisory council, was one of the issues for which he stated his government needed a mandate. With an election victory in hand, Frost would be well positioned for a showdown with the OMEA over the existence of the council. One did not occur, however. Instead, following a resolution at the association's February 1952 convention, the government agreed to recognize the OMEA as the body from which the HEPC should receive advice.[35] The provision for the council would remain in the Power Commission Act until 1973, but the council never functioned.

Many of the challenges that Frost had foreseen as requiring outside advice confronted the HEPC in 1952. In that year, three avenues for new energy supplies were opened in Ontario. The International Joint Commission approved the development of the St Lawrence after the myriad of provincial, federal, and international problems which had hamstrung the project since the 1920s had been resolved. Ontario followed by passing enabling legislation for the HEPC's participation on 23 October.[36] Then, on the basis of discussions between the Atomic Energy Control Board of Canada (proprietor of Canada's nuclear technology) and the HEPC, a joint effort began to develop nuclear power on a commercial basis in Ontario by 1962. Indeed, in anticipation of commercialization, Atomic Energy of Canada Limited was established as a corporate subsidiary of the AECB in 1952.[37] Finally, discussions began for the construction of a pipeline to transport natural gas to Ontario from Alberta to meet increased commercial and consumer demand. This would represent new competition for the HEPC.

Following the introduction of these mega-projects, Frost injected a measure of government input into the HEPC's affairs. In 1953 he created a committee on which senior treasury officials joined commission planners 'to coordinate the raising of capital and to consider the implications of Hydro's plans from the standpoint of general public policy.'[38]

Having previously been frustrated by both Saunders and the OMEA, Frost received an unexpected opportunity to revamp the HEPC's decision-making in 1955. Saunders, who was only fifty-one, died on 16 January from injuries sustained in a plane crash. Rather than have Challies, as vice-chairman, automatically become acting chairman, Frost refrained from making a minister chairman. Instead, he appointed Richard Hearn on an interim basis on 24 January. Unlike Hogg before him, Hearn relinquished his positions as general manager and chief engineer, maintaining the division of labour between chief executive and chief operating officer. At the OMEA convention that winter, the delegates used the occasion of Hearn's appointment to renew their call for direct and increased representation in an enlarged HEPC. The resolution, after having been moderated through amendment, respected the cabinet's prerogative to make appointments and only called for the government to consult it in the case of vacancies or an increase in the number of commissioners.[39]

The government responded to the OMEA's restlessness with a number of amendments to the Power Commission Act. When introducing the legislation, Frost stated that the size of the HEPC would be enlarged to recognize the OMEA's longstanding claim for further representation and to broaden input into the commission. However, Frost side-stepped the OMEA executive's request that he make appointments from persons whom the association nominated. Instead, he announced that the OMEA president, Lieutenant-Colonel A.A. Kennedy, would be appointed when the legislation had passed. He also stated that he wanted an executive committee of three concerned with daily management and the additional commissioners to function like directors of a private corporation.[40] This outline was somewhat similar to the division of labour Frost had intended for the commissioners and the advisory council, but it offered more collegiality than had been the case in 1951.

The amendments to the Power Commission Act, passed on 31 March 1955, had four main features: the number of commissioners was expanded to six from three; the number of vice-chairmen was increased to two from one (presumably so both the government and the OMEA could have one, given that Strike had informally been referred to as second vice-chairman since 1947); the chairman and the two vice-chairmen were designated the chief executive officers, able to exercise all the powers of the HEPC as an executive committee; and the $45,000 cap on overall remuneration for the commissioners as a group was removed.[41] As a result of these changes, Walter Gordon's 1947 recommendation that the commissioners concern themselves with policy while the general manager look after daily management was put to rest. In fact, the relentless effort to expand supply for postwar growth, the continuing frequency standardization program, and the statutory supervision of municipal hydro matters were onerous responsibilities for the commissioners. Not surprisingly, it was at this time that Strike gave up his law practice to work full-time at the HEPC.

After putting in place these structural changes, Frost proceeded to infuse the commission with new appointments, although not with the full complement of six commissioners. He created a vacancy by appointing Challies as chairman of the new St Lawrence Development Commission, occasioning Challies's resignation from both the HEPC and the cabinet. On 2 May 1955, with Hearn remaining chairman and Strike the OMEA representative, Frost appointed William Hamilton, a minister without portfolio, as first vice-chairman and OMEA president Kennedy, as promised, as a fourth commissioner. Hamilton, however, was defeated in the June election in which Frost won another majority. He was replaced as first vice-chairman by W.K. Warrender, another minister without portfolio, on 17 August. In making these particular appointments, Frost had in fact diverged from the Ferguson formula of balanced representation for the government and OMEA. Although their representation on the executive committee was equal, the government had just one representative on the commission, while the OMEA's representation was increased to two.

While the OMEA was pleased by its increased representation, many of its members had designs on the chairmanship. Hearn was set to retire on 31 October 1956, and they felt that Strike should be appointed chairman in his place. This was not to be. Rather, Frost appointed sixty-three-year-old James Duncan, the former chairman and president of Massey-Ferguson. Although Duncan was a virtual unknown to the hydro community, he had the advantage of having been friends with C.D. Howe, the federal industry minister responsible for AECL, since being federal deputy minister of air during the war. Thus, even though Duncan, according to Peter Cook, had been 'unceremoniously removed' from Massey, Roger Graham reports that Frost respected him highly and was glad he accepted.[42] As for Strike's view of the outcome, he recognized, according to his son, that he himself was 'not high profile or political enough,' but felt that Duncan was not an attentive chairman. From the staff perspective, Douglas Gordon, then director of consumer service, believes that Duncan simply gave a lot of control back to management.

In addition to Duncan's appointment, Frost made three other changes to the HEPC's membership. After Warrender became municipal affairs minister, Ray Connell replaced him as the minister without portfolio on the commission. Strike replaced Warrender as first vice-chairman, likely as a consolation for the OMEA, and Connell became the second vice-chairman. And Bud Cliff, the secretary-treasurer of the OMEA since 1953 and formerly its president, was appointed as a fifth commissioner, and possibly also as a consolation for not giving Strike the chairmanship. As a result, the OMEA now had three former presidents – Strike, Kennedy, and Cliff – as commissioners.

In sum, Frost had stuck to the Ferguson formula by appointing a non-partisan from outside the HEPC as chairman but obviously felt it was not essential to keep government constituency representation in balance with OMEA representation. Not surprisingly, the delegates at the February 1957 OMEA convention expressed their complete satisfaction with the changes initiated by Frost.[43] Although they had good reason to be jubilant that they now had three of the five commissioners, the responsibilities of the commissioners were differentiated by function. Thus, according to his son Don, Cliff worked only one day a week at the HEPC as an ordinary commissioner, and presumably the same held for Kennedy who lived in Owen Sound. As a result, balanced constituency representation remained on the executive committee through Strike and Connell. However, even here the situation had changed. Duncan, unlike Saunders who had been mayor of Toronto, had no hydro experience. The same held for Connell. Moreover,

Duncan travelled out of the country frequently and extensively, leaving Strike, in whom he had confidence, to take charge of the commission's affairs.

Among the first concerns of the new HEPC leadership was direct competition from the natural gas industry. The pipeline from western Canada was set to be completed in 1958, causing gas distributors to seek new customers actively in 1957. Until this time, residential space heating had not been of interest to the commission, but the promotional rates being offered to entice home owners to switch to natural gas changed the situation. It was not only advantageous to convert furnaces to gas; the same held for ranges and water heaters as well. Without these two important sources of electric demand, fear grew that the historical basis for low rates – high consumption – would be lost. To meet the challenge, the HEPC, in conjunction with the OMEA, devised the 'Live Better Electrically' campaign. Its details dominated the 1958 OMEA convention. For instance, Strike implored the delegates to recognize the self-interest in promoting, not just selling, electricity and the need to form a joint strategy with appliance manufacturers, dealers, and electrical contractors. The signal that the competition would hurt came when Duncan announced following the convention that the joint strategy would be extended to direct competition on residential space heating.[44]

Asserting Government Control

The relationship between the government and the HEPC changed dramatically in May 1958. After Connell was elevated to minister of reform institutions, Frost appointed Robert Macaulay, a dynamic thirty-seven-year-old minister without portfolio, as the new second vice-chairman. In Macaulay's view, the reason he was selected over other promising candidates was that Frost recognized he had shown initiative on the subject in his seven years in the legislature. Specifically, he had used his legal talents to redraft legislation to permit expropriations for the vast St Lawrence project and had become well versed in hydro matters. When Macaulay asked for the premier's view on what the job entailed, Frost responded that the HEPC was 'the largest god damn refrigerator in North America and I want to warm it up.' George Gathercole, then deputy minister of economics who would be appointed to the HEPC himself in 1961, concurs. In his view, Macaulay was appointed because Frost was seeking 'to know first-hand what was going on.'

Macaulay went about becoming an activist commissioner much to the consternation of Duncan, who, in Macaulay's view, did not want any rival as the 'connecting link' to Frost. As a result, Macaulay feels Duncan sought to 'breast feed' him information, whereas he was asking questions directly of the HEPC's staff. At first Duncan responded with bewilderment. In conversation with the provincial treasurer, Dana Porter (as later reported to Macaulay by Porter), Duncan said: 'We gave him a car, a chauffeur, we put a refrigerator in his office. What more could he ask for?' Believing Macaulay's inquisitiveness was without precedent, Duncan's bewilderment later turned to resentment. He made the case to Frost in August 1958 that Macaulay's probing amounted to partisan interference in the HEPC's operations, but the premier was unsympathetic.[45] Frost told Duncan, Macaulay recollects, that the minister was just 'a young man who is trying to fulfil his relationship between Hydro and members of the legislature and me.' Macaulay's appointment was nonetheless a watershed. According to Douglas Gordon, Macaulay's active interest and participation in what was going on inside the commission was what distinguished him from his predecessors.

The differing views of Duncan and Macaulay were aired publicly in the fall of 1958 before the Committee on the Organization of Government in Ontario. Chaired by Walter Gordon, the committee was mandated by Frost to examine 'administrative and executive problems' as well as the 'relationship of Boards and Commissions to the Government and the Legislature.' Moreover, Frost specifically mentioned the HEPC as a case in point of how the problems of governing have been greatly magnified. In the commission's submission, Duncan forwarded the view of the 1947 management consultant's report, authored by none other than Walter Gordon, that the chairman should report directly to the premier. Duncan also stipulated that he, as chairman, was anxious for MPPs and ministers to bring their hydro problems directly to him. This was a veiled critique of the liaison function performed by Macaulay. For his part, Macaulay recalls expressing an entirely contradictory view. He argued that a minister 'should have responsibility for the HEPC not just in reporting, but that [the minister] should be part of the [decision-making] process.'[46]

The contesting views of Duncan and Macaulay escalated further after the speech from the throne on 27 January 1959 announced that an energy portfolio would be created. The portfolio's stated purpose was to deal 'comprehensively with energy and power' in all its forms, 'determine the most economic use of power sources,' and 'deal with the vast financial problems in such developments.'[47] Although the announcement pre-empted the Gordon Committee's report, according to Macaulay it was not meant to prejudge the committee's recommendations. He recollects that Frost believed

Ontario was on the steps of a new generation in terms of energy and for this reason felt it was time to specialize, even if an energy portfolio would not please Duncan or the OMEA. Gathercole, who believes the opponents were unnecessarily perturbed, similarly recalls that energy policy issues simply had expanded to such an extent that a separate department was thought necessary if the government was to be able to deal with the questions for which it would be responsible.

Strike, for his part, told Macaulay and Frost that making the HEPC responsible to a minister with portfolio would be its death knell. According to his son, he felt the commission could not be efficiently managed if it had to kowtow to politicians. Until this point, he had accepted but not enjoyed the politics which ministers without portfolio had brought to the HEPC, choosing to object only when they went overboard. Strike nonetheless recognized that Macaulay differed from his predecessors because he had been appointed to make the commission more responsible to the government and the public. For this reason, Strike respected him as a role player for Frost and did not stand in his way.

In reply to the throne speech on 10 February, Macaulay offered a duplicitous account of the HEPC's origin and history. Anticipating the OMEA's objections to the proposed energy department, he went to great lengths to state that the municipal hydros were the owners of the HEPC rather than the government. But after acknowledging the OMEA as the representative of the owners, he implicitly criticized as 'untutored' the association's call to keep politics, rather than just patronage, out of the commission's affairs because of the government's legitimate interest as its creditor. In closing, and still without having mentioned the proposed department, he quoted Disraeli that 'change is inevitable in a progressive country.'[48]

Duncan was likely not impressed by Macaulay's failure to provide information about the proposed department, believing the department was being created specifically to control the HEPC. He expressed his apprehension directly to Frost on 17 February, offering two reasons why the commission should be excluded from its purview. First, he did not feel that Macaulay, the rumoured minister, would be 'content merely to provide a linkage,' as had the preceding ministers without portfolio. Secondly, he feared that the existence of the portfolio would threaten his direct access to Frost. He also expressed concern over a resurrection of the advisory council, although Macaulay claims that no plan existed. After introducing the legislation to create the department on 11 March, Frost sought to allay Duncan's fears by informing him that the department was intended only to develop broad energy policy.[49]

The government's decision to create an energy department was a call to arms for the OMEA. At its March 1959 convention, President Bert Merson reported that he and Secretary-Treasurer Cliff, who doubled as a HEPC commissioner, had expressed to Frost the OMEA's strong exception to the proposed department. Although Frost had assured them that it would not change the status of the relationship between the commission and the government, Merson sponsored a resolution calling on the government not to enact legislation which would 'circumvent the present independence of action' of the HEPC. Cliff, for his part, used his secretary-treasurer's report obtusely to express the OMEA's historic scepticism toward government control, stating that it was time to resurrect the 'spirit of cooperation' in the commission's origin. Duncan and Strike, however, sought to diffuse the situation. Duncan now claimed that he 'looked forward to the future without concern' and justified the department's creation on the basis that 'the government and all members of the legislative assembly are intimately affected by Ontario Hydro.' Strike deflated the OMEA's protests by reminding the delegates that public hydro had only been successful and efficient when the HEPC, the municipal hydro commissions, and the government performed their statutorily assigned functions with the utmost 'diligence, goodwill and cooperation.'[50]

The Department of Energy Resources was created on 26 March 1959. Macaulay, who drafted the statute, recollects that he was careful not to make the minister formally responsible for the HEPC in order to maintain the illusion of autonomy. And although it gave the cabinet the power to assign to the minister the administration of any other act, the Power Commission Act was neither assigned nor itself amended to make the HEPC responsible through the minister.[51] In fact, by statute the commission still submitted its annual report to the provincial secretary. According to Gathercole, there were two reasons the minister was not made officially responsible for the HEPC: 'too much government interference would create dissension,' and the minister, who presumably would fill the statutory requirement for a minister on the HEPC, 'had only one vote.' Nevertheless, Macaulay believes that even though the energy resources minister was not made formally responsible for the HEPC, it was generally understood that the minister would be responsible.

On another score, the Crown Agencies Act, which had been passed three days earlier in the session, had further clarified the HEPC's autonomy. Although it was introduced for the innocuous purpose of declaring all agencies 'owned, controlled or operated' by the government to be agents of the crown and thus exempt from federal taxation, it had been amended before being passed to exclude the HEPC from this designation.[52] While this amendment was presumably passed in reaction to his bill, Macaulay remarks that legislators in Ontario were almost working with abacuses at this time, causing most legislation to be introduced rather ineptly and many amendments to be knee-jerk reactions.

The Department of Energy Resources, at the time unique in the English-speaking democracies, was ostensibly created to permit a minister to address big-picture energy issues rather than assume direct responsibility for the HEPC. Given the greater diversity of energy supply and the increased importance of energy to the economy, it was necessary, Gathercole remembers, for the government to address the 'broader societal interest.' The basis for this general mandate, according to Macaulay, was that Frost felt a minister was needed to guide Ontario through its increasing reliance on thermal power generated from American coal. This was a politically sensitive matter, given the historic significance of inexpensive water power to Ontario's economic self-reliance. Although the government felt that the problem would eventually be cushioned by the use of nuclear power, Frost also believed that a minister was needed to find alternate uses for Ontario uranium since future demand for mining output from Elliot Lake looked increasingly weak. This was the case despite the fact that the decision to build a commercial nuclear power station at Douglas Point was made in 1959. In addition, there was the thorny issue of competition between electricity and natural gas to consume the minister's time. Donald C. MacDonald, the then Ontario CCF leader, writes that the creation of the department was an 'implied confession of inadequate policies' for the natural gas industry.[53]

Macaulay, as expected, was appointed minister of energy resources on 5 May 1959, one month before Frost won another majority. Despite this new responsibility, he remained the HEPC's second vice-chairman. This dual responsibility and the fact that the relationship between the commission and the minister was not mentioned in the statute was problematic for the Gordon Committee. Although a minister who sat as a commissioner was ostensibly subordinate to the collective will of the commissioners, by constitutional convention he had the power to prevail in policy disputes. Reporting on 25 September, the committee recommended that the minister carry out his responsibilities with 'care, foresight and understanding' to ensure that the HEPC's essential independence was maintained and usefulness left unimpaired. More specifically, it recommended that ministerial involvement in the commission's affairs should be limited to 'plans for growth and development and the more important of its operating policies.' Alternately, if the minister assumed responsibility for day-to-day administration the chairman and commissioners would possibly be left redundant. Moreover, the buffer they provided in terms of shielding the HEPC's operation from partisan control would be undermined. In the committee's view, the division of responsibility would best be safeguarded if the minister no longer sat as a commissioner because this practice presented confusing lines of authority.[54]

Although Macaulay remained both a minister and a commissioner, he believes the government reserved judgment on the Gordon Committee's recommendations regarding his dual role rather than rejecting them outright. He explains that the recommendations were a 'very modern view' of relations between a government and a crown corporation and that Gordon, whom he knew personally, was a man 'ahead of his time.' In retrospect, he believes the recommendations were correct but were not implemented immediately for two reasons: there were political overtones this message would have sent the OMEA; and Frost, who did not make rash moves, would not act on the recommendations until he had fitted them into his long-term plans for restructuring the HEPC. In creating the Department of Energy Resources, however, the government went in the opposite direction to the Gordon report. It recommended departments be consolidated in number and balanced in terms of size and responsibility.[55]

By having the dual role of energy resources minister and HEPC commissioner, Macaulay faced great potential for policy conflicts. He claims one never occurred, however, mostly because his presumed power over the commission was an illusion. Any potential for conflict was diminished by the fact that he did not attend meetings of the HEPC on a regular basis after assuming the portfolio. Furthermore, in his view the chairman held all the power, and Strike, the first vice-chairman, had important influence stemming from his municipal constituency which he as a government representative could not equal. Nevertheless, Douglas Gordon, from a management perspective, remarks that at the interface of the government and the HEPC, which historically was the function of the government representative, Macaulay was a strong minister.

Although the relationship between the government and the HEPC was not raised in the 1960 legislative session, Macaulay set about to consolidate the government's statutory powers in the energy policy field under his energy resources portfolio. He did this primarily through the Energy Act and the Ontario Energy Board Act. Although the latter transferred regulatory power over the oil and gas industry from the mines portfolio to energy resources, it did not extend to include the HEPC's rates.[56] Thus, the initiative was sufficiently restrained that the relationship between the government and the commission escaped scrutiny at the 1960 OMEA convention.

Later in the year, however, the HEPC's relationship with the government was scrutinized by a select committee of the legislature reviewing the Gordon Committee recommendations. In its 17 November report, the select committee argued, generally, for greater legislative scrutiny of agencies. In particular, it recommended that their annual reports be submitted to a responsible minister who would be answerable for the agency in the legislature. The committee specifically acknowledged the HEPC's 'cooperative trust' foundations but concluded that it was 'ultimately responsible to the Legislature.' The committee wished to undertake further study before making any specific recommendations on the HEPC, but none was undertaken.[57]

Following on the Gordon Committee's and the select committee's sentiments, the relationship between the HEPC and the government became an issue during the 1961 legislative session. This was occasioned by Bill 53 as initiated by Ross Whicher, an opposition Liberal MPP. Although he was inspired by the recommendations of the two committees, he had not been on the select committee. The Whicher bill proposed to make the minister of energy resources responsible for the HEPC, to remove the provision for ministers to sit as commissioners, and to appoint the provincial auditor as the commission's auditor, abolishing its prerogative to choose its own. Given the bill's controversial nature, it was much discussed at the OMEA convention in February. In arguing against the bill, President V.S. Wilson repeated the refrain that the HEPC was the trustee of a municipal cooperative rather than an agency of the government. The convention then gave its support to a motion sponsored by Bert Merson. Overlooking the fact that the bill was introduced by the opposition, Merson, in a misdirection of hostility, called for no action to be taken by the government. Conspicuous by his silence was Bud Cliff, who had not hesitated to give his view in 1959.[58]

When the convention was nearing an end, Macaulay arrived to make remarks on the Whicher bill, having received Wilson's permission and Frost's consent. From the floor, Merson strenuously objected to the entry of politics into the proceedings and challenged Wilson's invitation for Macaulay to address the convention. A standing vote had to be held to decide the issue, occasioning one of the most tense moments in the OMEA's history. When the delegates voted to sustain Macaulay's invitation to speak, fearing the consequences of spurning him, Merson and others stormed out of the room. Following this high drama, after referring to himself as the responsible minister, Macaulay stated that the government would keep Frost's assurance not to tamper with the traditional relationship. He then criticized the Whicher bill as designed to obliterate this relationship and assured the delegates that the bill would be defeated by the government. For this news, he received a standing ovation. In the end, the Whicher bill was defeated at second reading on 28 March.[59]

The 1961 convention also contained a portent of the future. Duncan felt it necessary to rebut criticism that the HEPC was engaged in a 'misguided effort to stimulate growth for growth's sake.' In his view, the criticism failed to distinguish the commission's sales promotion campaign, which was geared to promoting greater economy and efficiency from the existing capital works, from its new facilities which were required for the expanding economy. He then pointed out that in a residence with a gas furnace, range, and water heater, the revenue from electric use, primarily lighting, was not sufficient to offset the capital cost of the service. Moreover, he noted that growth rates in many municipalities were not sufficient to keep rates low, making it imperative to reverse the trend to gas.[60] Gordon McHenry, who was then the HEPC's manager of residential sales, explains that space heating was encouraged not only to retain the 'heavy' loads, but also because it had the potential to flatten the load curve by increasing power use in off-peak hours and thereby increasing revenue from a base of relatively fixed costs.

A Changing of the Guard

Both the premier and the HEPC chairman resigned in 1961, occasioning numerous changes to the actors in the commission's relationship with the government. Duncan resigned on 31 May at age sixty-eight and was succeeded as chairman by Strike, despite his age of sixty-five. Although there had been rumours that Gathercole would be appointed chairman, both Macaulay and Gathercole remember that Frost clearly wanted to appoint his old friend Strike as a reward for service. In turn, Macaulay replaced Strike as first vice-chairman without a new second vice-chairman being named. Strike's appointment was notable in that it was the first that did not fit the Ferguson formula since that of J.R. Cooke from 1930 to 1934. Even though he was not a partisan, he had been an OMEA constituency representative. Later in the year Frost stepped down and was succeeded as premier by John Robarts, his minister of education, on 8 November.

Hydro issues had played a small but not insignificant role in the transition of the Conservative party leadership and the premiership. Macaulay, who finished third in the leadership race, believes his views on hydro issues had an effect on the outcome since he was perceived by the OMEA as a threat, and its membership reached every Conservative constituency association in the province. Although there was no organized campaign against Macaulay, Bill Davis, who was then his campaign manager and later premier of Ontario, remembers that the OMEA presence affected but did not determine the outcome. In any event, after Macaulay had swung his support to Robarts on the final ballot of the convention, Robarts acknowledged his interests in energy and the economy. The new premier appointed Macaulay minister of commerce and development on 8 November, with the portfolio re-enacted as economics and development the next month. He also continued as minister of energy resources and, for the time being, the HEPC's first vice-chairman.

The subject of cabinet ministers sitting as commissioners was again raised when Macaulay introduced Bill 36 in the legislature on 1 December. Although he had condemned the intent of the Whicher bill earlier in the year, Macaulay was now acting on one of its key tenets. He was seeking to end the requirement that one commissioner be a minister, leaving the provision that two of the six may be ministers if so desired.[61] While this appeared to be a change of policy from the Frost to the Robarts government, Macaulay suggests it was a belated response to the 1959 Gordon report. A minister was apparently no longer required to be enmeshed in the HEPC's administration, although the provision was as old as the commission itself.

Following this diminution of the government's direct influence in the HEPC, Robarts made two appointments on 15 December 1961 that ensured there would be continued policy sensitivity for the government. For the first time the HEPC had the full complement of six commissioners that Frost had provided for in 1955. One of the two appointees was Gathercole, now deputy minister of economics and development under Macaulay. He was made first vice-chairman in place of Macaulay, who became a regular commissioner. Although Gathercole had begun his career in 1939 as economic executive assistant to the HEPC commissioners, his appointment to the commission was a first for a career civil servant. Macaulay recollects that the appointment was made by Robarts to honour a commitment to Frost. In return for his many years of service as an economic adviser, Frost wanted Gathercole, who was only fifty-two, to become HEPC chairman after Strike. Robarts, who had an equally high regard for Gathercole, agreed to the appointment as an interim step on the understanding, which later caused some problems, that Strike would stay in Macaulay's words for a 'short time' and in Gathercole's words for a 'few years.'

Gathercole's appointment was also based on merit because of his intimate knowledge of both the Ontario government and the Ontario economy. His background made him the embodiment of the general phenomenon of modernization within the government of Ontario. Having been provincial statistician, provincial economist, and the first deputy minister of economics since joining the government after the war, he had been, in Macaulay's opinion, the first to bring economic philosophy and a longer view to the government. Moreover, he had also risen to the top through the good judgment of his common sense as a policy adviser. In this regard, Ian Macdonald, who later became provincial economist and deputy treasurer, considers Gathercole an outstanding individual who rose above others as a man of action who accepted difficult tasks and achieved results.

Since the economic consequences of the commission were so great, a person of Gathercole's stature and capabilities was needed to bridge the HEPC's and the government's policy planning more closely together. Although Gathercole does not believe his appointment was motivated by any specific purpose, the most obvious being to coordinate the HEPC's growth with the government's economic objectives, such explicit direction was not necessary according to Macdonald because Gathercole knew where the HEPC fitted into the economy in both macro and micro senses. Moreover, with the government looking to the commission to be one of the principal instruments in shaping Ontario's economic growth, Gathercole was the person to provide the leadership because he had drawn up the government's economic forecasts as deputy minister of economics. While the objective was to coordinate rather than control and thus assure that the two did not work at cross purposes, Macdonald believes the government's purpose was also to maximize the economic benefits.

Macaulay, however, does not believe Gathercole was appointed to coordinate the HEPC's expansion with the government's economic objectives, let alone maximize the benefits. He suggests that view reads too much into the government's motivations. Having discussed economic issues with Gathercole while minister of economics and development and sat with him as a HEPC commissioner, Macaulay does not recall any grand scheme behind the appointment. Moreover, he does not believe that Gathercole was successful even if there were such a scheme because the commission, as an organization, was controlled from the middle rather than managed from the top. He allows that Gathercole might have been appointed to temper the HEPC's attitude that what was good for the commission was good for Ontario. However, even on this score, Macaulay does not think Gathercole had much success. Gathercole, for his part, does not even concede that he was appointed to speak for the broader societal interest, let alone coordinate the government's and the HEPC's planning.

Strike perceived Gathercole as the 'government man,' appointed to finish what Macaulay had started. In McHenry's view, a continuing tension was readily apparent between the two men from the start. Having been at the first meeting of the HEPC that Gathercole attended, McHenry recalls that Gathercole, who was unknown to most in the room, proceeded to take over the meeting much to the consternation of Strike. Tempering these views, Douglas Gordon, then the HEPC's executive director of marketing and later general manager under Gathercole's chairmanship, remarks that the government, in appointing Gathercole, simply selected a strong representative who would ensure a close relationship and better communication.

The second new appointment to the HEPC made by Robarts on 15 December 1961 was Bill Davis, then a promising thirty-two-year-old backbencher who had first been elected in 1959. Having supported Macaulay's leadership bid, he was appointed second vice-chairman on Macaulay's recommendation. Macaulay felt Davis was a comer and thus wanted to get him started up the ladder as soon as possible. To this end, the appointment was significant in that it added $10,000 to Davis's $7,000 legislative income, permitting him to make politics a full-time profession.[62] While this seems logical given future events, Davis, who was not then well established as a lawyer, suggests that he simply did not have the financial wherewithal to stay in politics in the short-term without this greater measure of responsibility. Macaulay had actually wanted Robarts to make Davis a minister to facilitate his own departure from the commission as the cabinet representative, but Robarts felt Davis was too young. This decision meant Macaulay had to remain a commissioner at least until his bill to remove the requirement for ministerial representation had passed.

Once Macaulay had been joined on the HEPC by Gathercole and Davis, it appeared that Robarts had positioned the three government representatives to balance the OMEA's three representatives, Strike, Kennedy, and Cliff. Although this would have provided Macaulay with a great deal of influence on energy policy as minister of energy resources, he suggests no such motivation prompted the appointments. In fact, juggling two cabinet portfolios and his seat on the HEPC, he recalls that he was too busy to attend all meetings. Although Davis, for his part, believes that balancing the OMEA's presence was never an issue, he nonetheless remembers Macaulay being quite consistent in wanting more control over the HEPC. Indeed, Macaulay would regularly remark to his young protégé, while pointing out the window of his Queen's Park office to the HEPC building: 'Those guys think they run the government!' Whether or not the government's presence on the full commission needed bolstering, it was clear that the two commissioners with government connections, Gathercole and Davis, outnumbered Strike of the OMEA on the HEPC's executive committee.

On the role of government in the HEPC's affairs, differing perspectives did surface between Strike and Gathercole at the 1962 OMEA convention. Strike used his address to give an account of the motivations underlying the creation of the commission, arguing, somewhat mistakenly, that the 'government had no desire to get into the electrical business' and that members of both parties had to be persuaded to pass the legislation allowing the municipalities to pay for the full cost of the undertaking. In his view, this origin had produced two rigidly followed principles: that the commission must retain all revenues; and that it must, as a highly technical business, be able to operate as a 'body corporate' without political interference. For his part, Gathercole, in his first appearance before the OMEA, pledged his support for the 'historic relationship,' but with qualification. He felt that although the 'HEPC should be free of political pressures on its day to day operations, ... [it] must carry out policy as laid out by the government of the day.' He closed by stating that conflicts between the municipal hydro commissions and the government in their 'partnership in power' must be 'openly and freely discussed.'[63]

Following the convention two important pieces of legislation sponsored by Macaulay and affecting the relations of the HEPC, the OMEA, and the government were approved on 30 March. One involved a consolidation of the HEPC's operation, for financial purposes, into one system. Until this time, the commission's works in northern Ontario were operated separately in trust for the government and, in the case of the Lakehead area, some municipal hydro commissions.[64] Through consolidation, the properties were vested absolutely in the HEPC. As with the frequency standardization works, the commission owned the property in its own right, leaving the municipal hydros in southern Ontario unable to claim rightfully that the property was part of what they owned cooperatively. Macaulay recalls that this was only a 'convenient side-effect': the real purpose was to structure the HEPC in a 'neater package.'

The other piece of legislation was the bill Macaulay had introduced in December 1961 to remove the statutory requirement for ministers to sit as commissioners. Paradoxically, the OMEA expressed its virulent opposition to this change in the 'settled view' of relations between the government and the HEPC, failing to acknowledge that the bill actually decreased the potential for government influence. Not surprisingly, the Power Commission Act was amended as intended.[65] While the termination of the remaining departmental feature of the HEPC's hybrid organization left its institutionalized ambivalence solely that of a government corporation and the trustee of a municipal cooperative, Macaulay nonetheless continued as both the minister of energy resources and commissioner for another year and a half. Meanwhile, Gathercole stepped down as deputy minister of economics and development in 1962 to work full-time at the HEPC. Davis resigned from the commission in November 1962 on becoming minister of education, whereupon Robarts appointed Robert Boyer, a Conservative backbencher since 1955, as second vice-chairman.

In an effort to resist further government control over the HEPC, the OMEA decided to publicize its longstanding view that the commission was the trustee of a municipal cooperative. In February 1963 it published *The OMEA – Ontario Hydro, The People's Power – and Government*. Written by Dr Robert Hay, chairman of Kingston Public Utilities, this booklet contended that the cooperative foundation of the HEPC had been obscured with the passage of time and the increased size of its operation. Moreover, the expressed exclusion of the commission from the Crown Agency Act of 1959 was offered as evidence of its not being a government agency.[66] In Macaulay's view, the basis for the cooperative claim – municipal hydro contributions to the HEPC's debt sinking fund – was an anachronism that only meant residual ownership would fall to the various municipal hydros in the event of a trustee sale. He nevertheless recalls that it was a powerful contention at the time because three of the commissioners, including the chairman, were former OMEA presidents. And the HEPC's annual report still detailed the extent of each municipal hydro's sinking fund contributions. Strike, for his part, was sympathetic to the cooperative view but, according to his son, did not believe it himself. He thought of the HEPC as a government corporation that had strong input from municipal hydro commissions because they were the largest customer. Gordon recalls that he looked upon the HEPC as being 'owned by the people of Ontario,' with the municipal hydros as 'strong partners in the total enterprise' of generation, transmission, and distribution of power under 'public authority.' Gathercole, somewhat surprisingly, states that he accepted the cooperative contention, believing it did not undervalue the government's contribution.

The OMEA's booklet was accepted in the report of its new government legislation committee at its March 1963 convention. Although discussion had centred on divorcing the HEPC from the government, the municipal cooperative contention was put to rest for the time being by Premier

Robarts at the close of the convention. In an address to the delegates, the first for a premier since Drew in 1947, Robarts acknowledged that the OMEA view had merit, but he let it be known that he considered the municipal hydro commissions, the HEPC, and the government 'partners' in the common cause of 'serving the people of Ontario.' Then, after reminding the delegates that three of the six commissioners were former OMEA presidents and stating that the government would continue to contribute, he closed by assuring the delegates that 'nobody [was] going to wrest the cable from your hands.'[67] According to Gathercole, the premier's speech served to quieten the OMEA's unrest.

After meeting the OMEA challenge, the government did follow through on the Gordon report's recommendation that ministers not sit as HEPC commissioners. Although he was returned with the government in the September 1963 election, Macaulay resigned from his two ministerial portfolios and as a commissioner on 16 October for reasons of ill health. He had been hospitalized after having collapsed in March and had not attended a meeting of the HEPC since 23 January.[68] His replacement as minister of energy resources, Jack Simonett, was not appointed to the HEPC, and a sixth commissioner was not named. The energy portfolio also lost its high profile and status by no longer being connected to the department of economics and development through the same minister. On reflection, Macaulay believes the department of energy resources need not have ever been created. Unless the minister had been given 'real responsibility' for the HEPC, as was done in Quebec in 1962, he believes that what remained for the portfolio should have been handled by another minister. In fact, he handled energy issues, when they arose, under his more senior portfolio of economics and development.

6

Taming the Hydro-Electric Power Commission: The Age of Government Modernization, 1964–73

Despite Macaulay's departure as minister of energy resources in 1963, the effort to revamp the Hydro-Electric Power Commission gained momentum, ultimately leading to its re-creation as Ontario Hydro in 1973. The government not only established Hydro in the image of a modern crown corporation but also succeeded in supplanting the informality of its past relations with the HEPC with more formal relations with Hydro. Although this revamping occurred in concert with extensive governmental reform, it nonetheless proved to be a long and arduous process. By raising the prospect of greater governmental supervision of the HEPC's affairs, the initiative renewed the tug of war over the commission's institutionalized ambivalence as both a government corporation and the trustee of a municipal cooperative. Considerable political will was required for the government to meet its objective given this struggle with the OMEA, but even then the commission's ownership ambivalence was only attenuated rather than terminated.

The HEPC's changing role in the economy provided the stimulus for the government to bring the existing commission structure and governmental supervision into line with contemporary expectations. During the 1960s the character of the HEPC was transformed noticeably on two interrelated fronts. First, it was elevated from a first-generation to a second-generation crown corporation. In becoming an instrument of industrial strategy and counter-cyclical economic policy by the 1970s, the commission moved beyond the basic service it had represented since the 1920s through the nationalization of a whole economic sector. Secondly, the HEPC was altered once more from a facilitative to a nationalistic crown corporation. Through its adoption of nuclear power on a large scale it no longer acted as merely a passive agent of economic growth but assumed a lead role in promoting the Canadian nuclear industry and Ontario uranium over technology and resources from other jurisdictions.

In response to the HEPC's increased importance as an actor in the Ontario economy, the government instituted a modern corporate structure and governmental oversight regime. Its traditional instruments for supervising the commission – statutory enactments and executive appointments – were no longer considered suitable for the task. Statutes had been used with regularity, although in piecemeal fashion, until 1962, but had never successfully overhauled the HEPC. Appointments, for their part, were restrained by the Ferguson formula, which meant the government could not freely appoint outsiders with experience in large corporations, as was its desire. The resilient municipal cooperative myth, however, stood in the way of the reform initiative, with the ensuing battle between the government and the OMEA limiting the outcome. Although the statutory basis for HEPC's ownership ambivalence was able to survive unamended, the more vulnerable symbolic basis was severely diminished on four counts: the introduction of a large board of directors broke the semi-convention of the Ferguson formula; responsibility for the HEPC was assigned to a minister with portfolio for energy; external review displaced the OMEA's self-proclaimed role as watchdog; and annual reports no longer detailed each municipal hydro commission's equity interest.

Prelude to Reform

Never having gained an official role in the HEPC's affairs, the Department of Energy Resources was reconstituted as Energy and Resources Management on 24 March 1964. Its new broader mandate encompassed not only energy but water resources, principally by assuming responsibility for the Ontario Water Resources Commission. This left its new minister, Jack Simonett, at pains to point out that electricity was just one of his responsibilities. When the OMEA tested him on why natural gas was being promoted in public housing when electricity was 'owned by the people,' he responded simply that it was not his function to 'promote one energy form over another.' Given that the HEPC embarked on the largest expansion program in its history in 1964, with the Pickering nuclear station as the centrepiece and the government in full knowledge, Simonett's statement revealed that his portfolio was not in the lead on energy policy.[1]

Simonett's exchange with the OMEA nonetheless reflected a new attitude, one which would continuously challenge the strength of the association and the notion that the HEPC was a municipal cooperative. The first instance arose that March when a select committee on local government, chaired by Hollis Beckett, recommended that municipal councils take over the function of local hydro commissions, and then, despite OMEA protests, went further in March 1965 by advocating regional government. Both cut to the heart of the association's existence. Where the latter would undermine its strength in numbers, the former removed its elected base, leaving the cooperative contention without articulation. The challenge was made concrete in June when Murray Jones, the review commissioner on Ottawa local government, recommended regional government, with hydro as both an upper-tier and council function. Wary of political interference in hydro matters, specifically that revenues would be siphoned off for other municipal expenses, the 1966 OMEA convention would call on the minister of municipal affairs, J.W. Spooner, to reject Jones's recommendations.[2]

Besides galvanizing the OMEA into action politically, the Jones report led the association to modernize its organizational structure. In 1965 Bud Cliff, the secretary-general who doubled as a HEPC commissioner, stepped down after having held this part-time position since 1953 for only a small honorarium. He had run the OMEA, recalls Andrew Frame who later became its president, 'out of a suitcase' between his private business in Dundas, Ontario, and his office at the HEPC. In his place, Ed Nokes was hired as a full-time secretary-manager in December, and a permanent office was established in the vicinity of the HEPC's head office.[3] This change was timely because the OMEA would remain overwhelmed by the plethora of studies and reform initiatives that affected local and provincial hydro.

Change was also in store for the HEPC's decision-making when Robarts filled the vacancy for a commissioner which had existed since Macaulay had resigned in 1963. Diverging from the Ferguson formula, he appointed Ian McRae, the retired chairman of Canadian General Electric and founder of its nuclear department, on 10 February 1966.[4] Chairmen aside, McRae was the first appointee since Arthur Meighen in 1931 not to have come from either of the HEPC's two historic constituencies, the government and the OMEA. Although he brought knowledge of the electrical and nuclear manufacturing industries to the HEPC's decision-making, according to Douglas Gordon, then assistant general manager of marketing, he was most valued for his general corporate experience.

The HEPC's increasingly corporate-like operation, not just municipal reform, had implications for the OMEA's municipal cooperative contention. A new uniform wholesale rate formula for local commissions, which had been implemented on 1 January with a 6 per cent rate increase, raised controversy at the association's 1966 convention. Strike compared the formula's significance to frequency standardization, and it had a similar effect by removing another vestige of the municipal cooperative foundation. Although there had been an ever-increasing pooling of costs since the 1940s, the new scheme replaced the general costing principle that each municipal hydro should be charged its proportion of the HEPC's overall costs. This principle, which was as old as the commission itself, had come to be considered unworkable with the integration of the transmission network in the 1950s and the consolidation of systems in 1962. Although the new formula was accepted, there was some sympathy for the complaint of Frame, then a Burlington hydro commissioner, that the HEPC had railroaded it onto the municipal hydros.[5]

Along with uniform rates, the changes also affected the debt retirement scheme. A case that reached the Supreme Court of Canada in 1982 with implications for Ontario Hydro's ownership was the result. After moving the basis of the sinking fund payments from assets to outstanding debt, the HEPC established a system of return on equity and cost of return to account for past contributions. Nepean, which began as a hydro municipality in 1964, eventually objected. By paying out more in cost than it received in return, it claimed to be subsidizing power costs in older municipalities. This new scheme, in effect, counteracted the uniform rates to municipal hydros, causing retail rates to vary by as much as 10.5 per cent by a 1972 estimate. In striking down the practice, the court would refer to the accumulated sinking fund contributions as only a 'notional' equity in Ontario Hydro.[6]

Following the convention, Strike, now seventy, stepped down as chairman on 31 March and was succeeded by Gathercole, who had been chairman designate since 1961. Gathercole remembers that Strike, by staying on longer than the two or three years anticipated, knew he was preventing a transition in the office, although Gordon believes Strike stayed on more because he liked the job than wanted to block Gathercole. Indeed, according to his son Alan, Strike considered hydro politics a 'brotherhood,' and it was one where his 'humility' and strong, genuine human relations had been an important factor in maintaining harmony in the provincial-municipal hydro relationship. What led him finally to bow out was the trend toward public consultation, which he found foreign to his way of operating. This was especially true of his earlier view that the HEPC only needed advice from the OMEA, and his self-described management philosophy of 'noble despotism,' which had earned him great praise in the early postwar era. First appointed in 1944, Strike's twenty-two years made him the longest serving commissioner ever, his term lasting three years more than Adam Beck's.

By keeping Gathercole waiting in the wings so long, the transition in the chairmanship was not smooth. Keith Reynolds, then Robarts's deputy minister, remarks that Strike had been under great pressure to resign after Gathercole had high-handedly insisted that Robarts should act on Frost's commitment to make him chairman. Gathercole, for his part, recollects having decided to take this step because Robarts was 'inclined to procrastination.' Despite this high drama, Gathercole, now fifty-seven, had been groomed for the position. Indeed, being the first senior civil servant to be appointed chairman suggested he was a quasi-deputy minister. While Ian Macdonald, then deputy treasurer, acknowledges that he had lots of consultation with Gathercole, his problem was that the chairman continually expounded the view that municipal hydros were the HEPC's sum and substance. Gathercole's appointment nonetheless fitted the pattern of chairmen in the Ferguson formula in that he was a non-partisan who could ensure for the premier that policy sensitivity would be exercised in the HEPC's decision-making. Although he readily fulfilled this function, Gordon suggests that, rather than being a 'government man,' there was no doubt that Gathercole was out for the commission's interests. In answer to the seeming paradox of his allegiances, Reynolds remarks that Gathecole was a political animal who liked being a public figure and enjoyed the controversy which that status carried.

On Gathercole's appointment as chairman, Cliff replaced him as first vice-chairman and thereby renewed the OMEA's representation on the HEPC's executive committee. The significance of membership on this committee was not clear, however, given that it no longer functioned as the management committee Frost had anticipated, although Gordon recalls that its members worked full-time and consulted regularly. Cliff, in fact, retired from his private business on accepting the position of first vice-chairman. Whatever the merits of his elevation, the OMEA's representation on the full commission was reduced to two when a sixth commissioner was not appointed in place of Strike.

In his first speech to the OMEA as HEPC chairman in March 1967, Gathercole drew attention to the positive impact of the commission's capital spending in expanding Ontario's economy and he addressed municipal hydro reform. The former was not well received given that the association's power-costing committee, which had come into prominence with high rate increases in the 1960s, was concerned about the effect of expansion and inflation on rates. In response, past-president Robert Hay, the committee chair, proposed two resolutions which were carried: that the HEPC use its financial reserve to keep rate increases in the vicinity of 2.75 per cent for the next five years; and that local hydro commissions expect and plan for rate increases of 2.5 to 3.0 per cent for the same period. Although neither would reflect a realistic picture of the future, Hay's research, Frame recalls, was always accepted on faith. On municipal reform, Gathercole felt the primary motivation of both the HEPC and the OMEA should be the preservation of autonomous commissions. In a similar vein, J.E. Wilson, the association president, stated that the OMEA should recognize the inevitability of reform but 'not accept change for the sake of change.'[7] The implication was that the HEPC was not opposed to municipal consolidation and that the OMEA was warming to the concept.

While the Ontario Committee on Taxation was known to have a broad mandate, its August 1967 recommendations were nonetheless startling for the degree to which they would alter the provincial-municipal hydro relationship and put upward pressure on rates. The committee, chaired by Lancelot Smith, recommended that the government not guarantee the HEPC's debt; that the HEPC be subject to all taxation; and that the sinking fund payments by municipal hydro commissions be replaced by a fixed charge. The last would undermine the main statutory provision supporting the municipal cooperative contention.[8] On local government reform, it recommended that all revenue-earning enterprises, including hydro commissions, pay full taxes; that councils control their surpluses; that the Department of Municipal Affairs supervise their operation (replacing the HEPC for hydro); and that regional government be initiated province-wide.[9]

Although the Smith Committee recommendations did not draw the OMEA's immediate attention, not even in the October 1967 election which gave Robarts a second majority, Gathercole publicly condemned the fundamental nature of the proposed changes at the association's March 1968 convention.[10] On these criticisms, Ian Macdonald remembers that Gathercole was too 'politically sensitive and astute' to be speaking only for himself. In general, he recalls that the government did not want to see consumer costs increase and thus did not favour the HEPC or local hydros paying full taxes even though this gave them an advantage over other energy producers. As for the removal of the HEPC's debt guarantee, he argues it was a non-starter for two reasons: it went against the trend of closer relations desired by the government, and would have made borrowing more expensive. With these aspects of the Smith report belittled, Gathercole turned his attention to explaining the 6 per cent rate increase the HEPC had instituted on 6 February, unable to keep it in the 2.75 range that the OMEA had recommended. While he stated that the commission faced the burden of a number of unfavourable but transitory factors, he also continued to propound the old refrain that the key to keeping rates low was to build diversity into load use, and thereby increase revenue from fixed costs. He did, however, expect 4 per cent increases for at least two or three years.[11]

With the February 1968 speech from the throne having announced that regional government would proceed in Ottawa, the issue was prominent at the convention. To the delegates' surprise, Darcy McKeough, the new minister of municipal affairs, had decided on the eve of the convention, after having been confronted by the OMEA, that the existing arrangements for hydro would not change. Looking back on the event, he recalls still favouring upper-tier hydro, but gave in because there were so many other battles to fight. In fact, when the government did incorporate regional government in its Design for Development planning framework later that year, hydro was not enumerated as a prospective upper-tier function.[12] But to McKeough's surprise this time, the OMEA executive supported upper-tier hydro in a 9 January 1969 brief as long as autonomous commissions remained. This accord ran into trouble when the association's convention that year took the executive to task for making policy without the consent of the membership. As part of a larger backlash against regional government, this reaction had some standing as Robarts instructed McKeough in May not to initiate any new regional government studies. In 1970 the OMEA would approve a compromise policy that was firm on autonomy from councils but flexible on the tier, with the latter to be determined by 'future economic conditions, economies of scale, and supply conditions.'[13]

On the other recurring and contentious issue, the HEPC had again increased rates on 1 January 1969 by 4.5 per cent for municipal hydros and much higher for its direct rural and industrial customers. With the demand for power continuing to rise, Gathercole was no longer just recommending that load be increased to keep rates down; he was now counting on the benefit of nuclear power to keep costs down. In the commission's annual report for 1968, he stated that large-scale nuclear plants could provide energy cheaper than fossil-fuelled plants and even hydro-electric plants located in remote areas. To this end, the Douglas Point nuclear generating station was now on stream, the Pickering station was under construction, and the Bruce station had been announced in December. For Gathercole, the additional allure of nuclear power was the 'advancement of a new industry in Canada with broad implications for the whole national economy.' On his nuclear power pronouncements, Gathercole recollects that he saw it as his role to match energy choices with what was good for the economy. In going nuclear, he remembers the economic spin-offs vis-à-vis fossil fuel plants as an important determinant, one which the government had encouraged.[14]

Despite its effect on rates, the OMEA came to the defence of expansion when the HEPC's plans faced public scepticism. Through a resolution which he knew would be quashed in the face of received wisdom, Frame, now a vice-president of the association, actually engineered mock dissent at its 1969 convention to illustrate the point. His purpose was to differentiate the electric industry from other energy producers, such as oil and gas, where demand could easily be controlled by restricting supply. Thus, after meekly calling for the HEPC to 'consider the advisability' of slowing down the rate of expansion to a more 'manageable' level, he proceeded to belittle the motion's intent. With competition still strong among energy suppliers, former president Wilson added, sardonically, that the motion should have been moved by Consumers Gas and seconded by Union Gas.[15]

The remainder of 1969 was relatively quiet, except for a number of changes to the actors in the HEPC's relationship with the government and Robarts's decision not to create a royal commission on HEPC rates. First, Simonett resigned from cabinet on 5 June and was replaced as minister of energy and resources management by George Kerr, to whom Frame was a friend, neighbour, and informal political adviser. Secondly, Cliff resigned from the HEPC on 30 June. His official reason was that at age sixty-eight he felt younger people should take over, but privately, according to his son, he chose to withdraw rather than concede to Gathercole in a financial policy dispute. Thirdly, Boyer, the Conservative backbencher and commissioner, replaced Cliff as first vice-chairman, and McRae replaced Boyer as second vice-chairman. Although Kennedy of the OMEA was the longest serving of any of the commissioners, the elevation of McRae, the one commissioner with corporate experience, meant that a representative of the association was no longer one of the chief executive officers. Finally, Robarts appointed Dr Douglas Flemming, the OMEA president in 1967–8, to the HEPC on 1 October 1969, after the association had requested that a new representative be appointed to replace Cliff. As for a royal commission on rates, Reynolds, now deputy minister to the premier and secretary to the cabinet, made the recommendation in November, but Robarts rejected it outright as handing ammunition to critics of the government.[16] Calls for rate regulation would be the outcome.

The Committee on Government Productivity and Task Force Hydro

Giving it the mandate to 'inquire into all matters pertaining to the management of the Government of Ontario,' Premier Robarts appointed the Committee on Government Productivity (COGP) on 23 December 1969. Chaired by John Cronyn, senior vice-president of corporate affairs for Labatt Breweries, the COGP divided its task into three projects: cabinet decision-making, department organization, and agencies, boards, and commissions. The agency project, whose focus was to be the 'organizational structure and reporting relationships best suited for the implementation of government policy,' raised the question of whether the HEPC's historic autonomy from government would be lessened or removed.[17] While Cronyn recollects that the extent of his committee's mandate was never very clearly pointed out, Reynolds, also a member of the committee, maintains that the HEPC was not intended to be included under its auspices. From the HEPC's perspective, Gathercole claims that he was neutral on the question and, like Frame of the OMEA, believed the commission would not be included. In time, however, the COGP would be instructed to scrutinize the HEPC, while the OMEA, preoccupied with the implementation of regional government, failed to ensure that the commission would escape scrutiny.

Meanwhile, Gathercole again proclaimed his faith in expansion in the HEPC's 1969 annual report, stating that there was no reason to suppose that demand would fall short of the long-term trend over the last fifty years. This was despite having increased rates by 6 per cent on 1 January 1970. He recognized that the new challenge was to keep supply and demand in balance, but saw the resolution of such problems in strengthened rate stabilization reserves, rather than in smaller-scale projects. Given the magnitude of the expansion, he continued to promote its economic benefits, which he declared to be the spin-offs of capital spending and, more generally, the commitment to Ontario's growing economy. He also noted that public pressure for the maintenance and improvement of the environment had increased costs greatly and, in particular, that the acquisition of property faced growing resistance. He revealed that the HEPC had allocated considerable financial and staff resources to the resolution of these concerns. Moreover, he was now claiming that the extension of electricity use was beneficial because it would be a factor in the abatement of pollution. Despite growing public concern for the environment, the OMEA remained oblivious to the changing public attitude, with its March convention again consumed by rates and municipal government reform.[18]

The HEPC lost its one commissioner with corporate experience when McRae died in September 1970. He was not replaced as second vice-chairman, a decision which ended the practice of having two vice-chairmen. With Boyer, now the only vice-chairman, dividing his time between an increasingly active legislature and the commission, only Gathercole was full-time. As a result, the executive committee of three chief executive officers begun in 1955 did not function, with Gathercole, for all intents and purposes, the CEO and the other commissioners as part-time directors. With no new appointments at this time, the HEPC had only four of a possible six commissioners. Besides Gathercole and Boyer, they were Kennedy and Flemming, both former OMEA presidents. By any measure, the commission lacked the degree and diversity of input enjoyed by other corporations its size.

Gathercole nonetheless sought to keep the HEPC responsive to public concern, expounding in its 1970 annual report that 'a "goods" society and a good society [were] not mutually exclusive.' He added that demand continued to increase and would double, as expected, within ten to twelve years despite a downturn in the economy and increased rates. In meeting this need, he repeated his new refrain that electric power would make the 'economy more efficient and the environment cleaner and healthier.' He also acknowledged, however, that the general trend of higher rates would be even more apparent. The reason was that thermal power, which surpassed 50 per cent of the HEPC's capacity for the first time in 1970, had high operating costs in fossil fuel plants and faced the 'problems inherent in the introduction of prototype equipment' in nuclear plants.[19]

As a result of the persistent large annual rate increases, rate regulation had become a burning political issue. Gordon, now the HEPC general manager, recollects that the call originated with the Niagara Basic Power Users, an interest group for large cost-sensitive industries that had regularly been consulted on rates but did not enjoy the status of the OMEA in the rate-making process. Given that the works supplying the Niagara area had been well paid for over the years, this group felt its members were subsidizing expansion elsewhere in the province because of the HEPC's commitment to uniform rates. With its specific objection, according to Gordon, being that the commission acted as judge and jury, the group had often made appeals directly to the cabinet. In addition to this group's complaints, consumers were beginning to make comparisons with American jurisdictions that had regulation, and some members of the legislature were disturbed by the size of the annual increases.

At the March 1971 OMEA convention regulation was perceived to be an assault on theassociation's watch-dog role. President D.G. Hugill argued that the call failed to recognize that the OMEA's power-costing committee kept rates under continuous study and scrutiny. Hay went further, suggesting that the call sought to undermine the nature of the HEPC as the trustee of a municipal cooperative. In his view, the existing procedure for setting rates – twelve interim monthly billings with a thirteenth bill for adjustments – was 'an essential mark of a true cooperative.' In a special symposium on the issue, A.J. Bowker, who became OMEA president in 1974, presented a report that ridiculed regulation as a desire for some 'magical way to curb rate increases.' In his view, the need for regulation was obviated by full disclosure, power at cost, and the HEPC's supervision of the municipal hydros. Although he was willing to accept that a formal public review of rates might be appropriate, he doubted, paradoxically, whether full regulation would 'ensure that the customers' interests [were] kept foremost.'[20]

In his address to the delegates, Gathercole tackled the regulation challenge with considerable equanimity. He responded to criticism of rate increases with the position that rates were still lower in Ontario than in most other North American jurisdictions. On the criticism of the HEPC's promotion of consumption, he added that rates would be substantially higher without new sales because of the need to spread the burden of fixed costs. In addition, he defended continued expansion on the presumption that electricity was 'destined to be a larger and larger share of the total energy market' in North America, and on the grounds that restricting its use would be 'environmentally self-defeating.' After acknowledging that 'freedom of action' had permitted the many achievements of hydro in Ontario, he urged the OMEA to recognize that decision-making must reflect the 'broad interests of the public and the province.'[21]

Although it had stymied almost every previous effort to revamp the HEPC, the OMEA was not prepared for the lightning pace of change after Bill Davis succeeded Robarts as premier on 1 March 1971. And since the transition coincided with the association's convention, its executive was unable to report whether Davis had any plans for the HEPC. While there was no shake up in the energy portfolio, where Kerr remained the minister, the portent for change arrived quickly when the COGP delivered its report on cabinet decision-making on 16 March. In place of the traditional departmentalized cabinet, it recommended an institutionalization of decision-making through a committee system and a formal policy process. Given this direction for reform, the COGP's yet to be forwarded recommendations on agencies, boards, and commissions would likely be sceptical of their autonomy, preferring that they be subject to government policy.[22]

The HEPC would escape the confines of the committee's specific recommendations, but not its scrutiny. The speech from the throne on 30 March explicitly requested the COGP to review the commission's 'function, structure, operation, financing and objectives.' The COGP, in turn, established Task Force Hydro as a 'semi-autonomous unit' in April.[23] The appointments to the task force, like the COGP itself, were by design a mix of business leaders and senior civil servants, with the HEPC, the OMEA, and large power users well represented. They were Dean Muncaster, president of Canadian Tire Corporation as chairman; Douglas Gordon, general manager of the HEPC; Andrew Frame, president of the OMEA, 1971–2; Robert Taylor, vice-president of the Steel Company of Canada (a member of the Niagara Basic Power Users); Keith Reynolds, secretary to the cabinet, deputy minister to the premier, and a COGP member; Hugh Crothers, president of Crothers Limited; and Richard Dillon, dean of engineering, University of Western Ontario, as a task force member and its full-time executive director. Reynolds, Dillon, and Taylor had been friends since their prewar days at the University of Western Ontario.

Despite this rapid sequence of events, the task force had not been created as a hurried reaction to the implications of the COGP's recommendations for the HEPC. Indeed, Cronyn recollects that Robarts had requested the committee to study the commission in 1970, long before Davis became premier. Cronyn and Reynolds, in turn, had paid visits to Gathercole to work out the logistics. Although Gathercole, according to Cronyn, was not at all keen on having the HEPC scrutinized, the purpose of the meetings, in Reynolds's words, was to inform him of 'how it was going to be.' Nonetheless, Cronyn, Gathercole, and Reynolds all maintain that the separate study was the committee's idea alone and not a compromise. Arguing instead that Gathercole used his clout to get a separate study, Dillon believes the hydro task force was created to 'soothe ruffled feathers' because the HEPC was 'jealously guarding its autonomy.' Macdonald, who was also a COGP member, remembers that the outcome was a 'marriage of convenience' in that Gathercole did not want the HEPC 'deemed an integral part of government' and Cronyn recognized it was 'too big to fight over.'

Despite the political intrigue, there were sound reasons for a separate study. The HEPC differed from other agencies in terms of not only its size but also its function and history. In Muncaster's view, the separate study was required because a number of new issues were coming to the fore, in addition to the large impact of the commission on the overall operation of the government and its highly specialized nature. Macdonald, on the other hand, believes that the motivating factor, in addition to size, was the HEPC's municipal connection rather than its science-based function.

Task Force Hydro turned out to be only the first step to ensuring Ontario's energy needs. Premier Davis created the Advisory Committee on

Energy as the next step on 22 July 1971. It was mandated to 'ascertain the future energy requirements and supplies of Ontario' and make recommendations towards meeting those requirements. Given apprehension over looming energy shortages and emerging debates over the limits of growth and nuclear power, the government was aware that the future would not look like the past. It felt that energy deserved a special look so policy making would not fall by default to the HEPC. Although this meant some overlap with the task force, the advisory committee's primary focus was the machinery of government rather than the delivery or HEPC side of the energy policy equation. Chaired by John Deutsch, the principal of Queen's University, it was given broad representation. In addition to all the major industry associations, the HEPC was directly represented by Gathercole; the government by, among others, Reynolds and Macdonald; and the hydro task force by Reynolds and Dillon. The Consumers' Association of Canada and Pollution Probe were also represented, but not the OMEA. When Frame wrote to Davis on 28 July to object,[24] McKeough, now treasurer of Ontario, responded that consumers were already represented. McKeough recollects that he snubbed the OMEA deliberately because the government was looking for 'fresh thinking,' which he knew would not come from this 'fossilized' group.

At the same time, the Department of Energy and Resources Management was collapsed into the Department of the Environment, a new portfolio created on 23 July, with Kerr remaining the minister. According to Reynolds, this department, which was established prior to the COGP's recommendations on departments, was created to consolidate related functions rather than to downgrade energy. Its creation would nevertheless be consistent with the committee's emphasis on fewer and larger departments. When viewed alongside the accompanying passage of the Environmental Protection Act, the government undoubtedly was seeking to improve its standing on environmental issues for the forthcoming election.[25] The government would be returned with a majority in October.

There were also a number of changes among the actors who articulated the HEPC's relationship with the government in 1971. Boyer, the vice-chairman and Conservative backbencher, had resigned from the HEPC on 30 June after announcing that he would not stand for re-election. Arthur Evans, a government backbencher who had been a local hydro commissioner and OMEA district director on his first election in 1960, was appointed vice-chairman in his place. With Kennedy and Flemming, both former OMEA presidents, and Gathercole the only other commissioners, two vacancies remained. Davis appointed Lou Danis, then a vice-president of the OMEA, and Roger Seguin, a lawyer and chairman of the board of governors of the University of Ottawa, as commissioners on 8 September. While Danis resigned from the OMEA executive on his appointment, the fact remained that three of the six commissioners were again former OMEA executives, and a fourth, Evans, had a strong familiarity with the association.

Instrument of Government Policy

In an economy facing instability, the Ontario government charted a new course in its relationship with the HEPC in 1971. It instructed that rate increases slated for 1 January 1972 be deferred to a later date in order to stimulate the economy. The decision was predicated, Davis recalls, on it 'being in the economic interest of Ontario, which in the long run was in the economic interest of Hydro.' For this reason, he places the event into the larger picture of the government's evolving capacity for economic forecasting, which he remembers indicated that the HEPC and the government should cooperate rather than operate in totally separate vacuums. According to Macdonald, under whose direction the initiative originated, the commission accepted its broader responsibility, albeit reluctantly. Indeed, Gathercole recalls that the HEPC felt that rates should rise, and Gordon recollects that electricity costs were such a small percentage of total costs for most companies that increases would not have a pronounced negative impact on the economy. Moreover, Dillon remarks that rates were already so low that intervention would not have had much effect. The government recognized these factors, Macdonald maintains, but proceeded nonetheless because the rate policy was just one component of a larger, interrelated response to the poor economic climate.

Although the OMEA's cooperation on the rate deferral was not required, its opposition consumed a great deal of Davis's time. The problem was that the rate intervention challenged the association's contention that the HEPC was the trustee of a municipal cooperative. Moreover, the OMEA had begun to take its view even more seriously with the onslaught of high rate increases. Indeed, following its 1971 convention, where calls for regulation had been denounced, the association had published *Power Costing in Ontario*, written by Hay and others to improve its own members' knowledge of the at-cost or cooperative rate-making formula. Despite having established this stake in the rate-setting process, the OMEA was not consulted on the government's intentions; it was only informed after its power-costing committee had set a meeting for 22 October, the day after the provincial election. While the campaign was in progress, Frame reveals, he was contacted by Davis's office after Gathercole had told Davis that 'you better wise him up.'[26]

Dissatisfied with what had transpired, the power-costing committee met with Gathercole on 26 October, and Frame, Hay, and Nokes, the secretary-manager, met with Davis on 5 November. At the latter meeting, the delegation agreed to support the government's rate intervention, and the HEPC formally announced the deferral in November. Before giving their consent, however, they expressed the OMEA's long-held view that, irrespective of economic conditions, rates should meet costs. Although Davis acknowledged their view, Hay later reported that the premier also questioned unrestricted growth in electricity usage and unlimited expansion of facilities. According to Frame, these limits to growth sentiments took the OMEA leaders by surprise. Thereafter, they used the meeting to broach the subject of their association's lack of representation on the advisory committee on energy. Davis was informed that Deutsch, the chairman, was willing to have the association included, and Davis responded by appointing Hay.[27]

While the OMEA objected to the government's rate intervention, its booklet on power costing had failed to point out that the government had always been consulted before rates were set. Indeed, Davis, on reflection, views the deferral as part of a 'continuation of communication' between the HEPC and the government and thus only an 'evolutionary' departure from past practice. Macdonald concedes that in this simpler time rates were set through the nexus of the premier and chairman, although only 'in a highly informal and off the record way.' Gordon acknowledges this fact but nonetheless views the rate deferral of 1972 as 'unprecedented' because the government had previously never exercised power of approval over rates. While Frame maintains that the OMEA knew that rate increases were monitored by the government, he argues that it never challenged this practice because the increases had usually been nominal. Dillon, on the other hand, believes that the OMEA, thinking it had been running the HEPC, was shaken by the reality of who had always determined rates. He adds that political considerations usually moderated rate increases at the expense of raising enough capital internally and hence at the expense of an appropriate debt/equity ratio.

The rate deferral was followed by another indication of increasing government influence in the HEPC's affairs after the COGP recommended in December 1971 that there be three policy ministers completely free of departmental obligations. The functions of these ministers would be to develop policy through the chairmanship of a policy committee of cabinet in conjunction with related line ministers in the respective fields of resources development, social development, and justice. In addition, the committee suggested that line ministers should be responsible for 'ministries,' adding to their current responsibilities all agencies, boards, and commissions related to their departments. Davis acted on the recommendation regarding policy ministers on 5 January 1972 by appointing past rivals for his party's leadership to the posts.[28] Regarding the HEPC, Bert Lawrence was appointed provincial secretary for resources development with Reynolds as his deputy minister, but the HEPC, given the lineage of the energy portfolio, was not made responsible to the minister of the environment.

On the fate of the expansion program in the recession, Gathercole himself was calling for the HEPC to be used for counter-cyclical economic policy. In its 1971 annual report, he noted that although energy growth had been moderate in comparison to years past, the construction program needed to be vigorously maintained because there was no reason to expect that growth would remain slow. In his view, he was arguing not for unrestrained expansion but rather to derive from electricity the 'maximum social benefit from the facilities to sustain Ontario's prosperity.' In making these pronouncements, Macdonald suggests that Gathercole was not acting on government-directed economic objectives, although he knew 'intuitively' which way to go. Gathercole's optimism nonetheless belied what he noted was a 'seemingly less tractable problem,' the disparity between the commission's revenues and costs. While this must have been aggravated by the rate deferral for 1972, Gathercole defended cooperation with the government to avoid blunting economic programs.[29] Only two months passed, however, before Davis informed Frame privately at the official opening of Pickering on 25 February 1972 that the deferral had been a success and that rates would soon be increased. According to Macdonald, other factors than just the rate deferral had been responsible, and although the slump was short, the recovery was 'quite noticeable and marked.'

The HEPC's 1971 annual report was also notable as the first not to detail each municipal hydro's sinking fund contributions. This practice had persisted even though the statutory requirement begun in 1918 had been dropped in 1950. In adopting a corporate-style format, the annual report continued to list only the total of these debt retirement payments on the balance sheet. Detailed information on each municipal hydro, however, was published thereafter in a new statistical yearbook that continued to list the contributions as 'equity in the HEPC.'[30]

Although Gathercole officially welcomed the reviews by Task Force Hydro and the Advisory Committee on Energy at the OMEA convention in February 1972, the issue he considered most pressing was rate regulation. Its immediacy sprang from the growing criticism that the HEPC 'operated as a law unto itself, set its rates arbitrarily and was answerable to nobody.' Labelling the criticism patently absurd, Gathercole argued in rebuttal that the HEPC was accountable to the municipal hydro commissions and through them to the people; that its commissioners served at the pleasure of the government; that a minister reported on it to the legislature; and that the government and the OMEA were 'consulted and advised' on power costs and rates. Given that the above involved continuing review, he stated there was no need for regulation, failing to recognize that there was no external scrutiny. Before closing, he announced that municipal rates would increase by 8 per cent on 1 July, an effective annual increase of 4 per cent, although rates would not increase for direct customers until 1 January 1973. He added that a municipal increase could be expected then as well and 'for many years to come.'[31]

In the power-costing committee's report to the convention, Hay argued that the nature of the HEPC as the trustee of a municipal cooperative was under siege. The rate deferral, in his view, indicated that the cooperative principle of at-cost power had been reduced to a mere obstacle in the face of overt political influence for a purely political exigency. In addition, he noted that the steep growth in the HEPC's borrowing was leading some observers – presumably the COGP and the task force – to question its autonomy from government, with the effect that the municipal hydro commissions' claimed equity ownership of the HEPC was in jeopardy. Moreover, he argued that the municipal hydros themselves had given public criticism of the commission credence by failing to realize that the public wanted the inclusion of environmental costs in the price of power.[32]

Hay's solution to the challenge was to give the OMEA's municipal cooperative contention new vitality. Towards this end, he proposed that a capital levy be included in the HEPC's wholesale power rate, arguing that it would still leave electricity a bargain at two or three times the price. In his view, a levy had numerous selling points: it would give the commission freedom from the 'bondage' of financial institutions and the 'political pressure' from the government guarantee of debt; it would end the criticism that the HEPC received an unfair advantage from this guarantee; it had the potential to make regulation unnecessary because it would give the commission the increased flexibility to 'adjust to environmental and social pressures'; and it would 'bring an end to the question of who owns Hydro.' Having heard Hay's report, the convention recorded its support for a capital levy and its opposition to rate manipulation.[33]

Outside the OMEA, the capital levy scheme was received with scepticism. Although the task force recognized that the HEPC had a poor debt/equity ratio, Muncaster recollects that he felt the levy scheme was inappropriate: it would have charged current users for future expansion; and it was unnecessarily rigid by requiring legislation. Furthermore, the 'political motivation' of the scheme would not have sat well with the financial community. McKeough, then treasurer of Ontario, remembers the levy scheme as 'pie in the sky,' which he doubted the HEPC supported. He was nonetheless unwilling to relinquish the government guarantee of the commission's debt because it was one of the few levers for insisting on consultation and thus was an important control mechanism. Gathercole recalls that the HEPC gave the levy consideration, but not in lieu of the government guarantee. The reason, Macdonald contends, was that the commission wanted the financial benefit of the guarantee even though it would have preferred to avoid the associated burden of discussion and collaboration with the government. Gordon recollects, however, that although the HEPC preferred to retain the guarantee, there were rumblings that it could afford to forego the guarantee and that the cost might be worth the trouble.

Given the OMEA's support for greater HEPC autonomy through the levy scheme, the delegates were disturbed when Muncaster, in an address to the convention, stated that the task force favoured the commission becoming more responsive to government policy. Although the government would be expected to articulate policy without interfering or intervening in the its day-to-day affairs, the HEPC was being considered for a host of roles as an instrument of government policy. The one he specifically rejected, however, was rate manipulation to stimulate the economy. In a view similar to the OMEA's, he stated that rates must cover total costs, including all social and environmental costs. Despite this basis for common agreement, the association was alarmed when he also remarked that the task force favoured a 'rationalization' of distribution.[34] Like regional government, this scheme would reduce the total number of local hydros and thus dilute the strength of the association. In retrospect, Muncaster feels the OMEA likely perceived rationalization as an attempt to undermine the association.

Despite the OMEA's sentiments, the HEPC fell under further government control through the Government Reorganization Act of April 1972. The act, besides enacting the portfolios of the policy ministers and changing all departments to 'ministries,' included an important change in terms of ministerial responsibility for the HEPC. It amended the Power Commission

Act to direct the commission's annual report to the minister of the environment,[35] now James Auld who replaced Kerr in February. This was the first time that the HEPC had ever been mandated by statute to report to a minister even though ministers had sat as commissioners from 1906 to 1963 and an energy portfolio had been created in 1959. While this undermined the myth of the HEPC's autonomy from government, Frame recollects that the OMEA did not protest because it missed the shift in the flurry of change. He adds, nevertheless, that the association had come to accept that the energy and later environment minister was already responsible for the HEPC.

Although making the HEPC report to the minister of the environment was consistent with the COGP's 'ministries' concept, this did not occur as a result of a specific or comprehensive committee recommendation. Cronyn recollects that the change occurred because the rapport was not good between Gathercole and Davis, to which Gathercole only remarks that their relations were cordial. Gathercole had had direct access to Robarts, but according to Cronyn, Davis wanted it made clear that the HEPC did not have 'a sort of God-given position in the universe.' Dillon remembers that Davis did not so much object to the old boy network of the premier and the chairman as he did not want to be bothered directly by the irritation at this time over HEPC's affairs. Acknowledging that the change was a departure, Davis maintains it did not result in response to any specific event like the rate deferral. In his view, the new reporting relationship, given the need for close relations on financial and other matters, was part of a natural evolution that made sense to both the HEPC and the government. In a similar vein, Macdonald contends that the change was part of a larger process designed to relieve pressure on the premier's office by leaving it a 'voice of last resort' that was out of operational discussion. The decision, McKeough reveals, was arrived at only after lengthy discussion in cabinet because the commission only wanted a relationship with the premier, if anybody.

On another front of change, Task Force Hydro's advocacy of a rationalization of municipal distribution was adopted in June 1972 by McKeough in the government's ongoing Design for Development program. McKeough explained that the delivery of services in a framework of a 'rational management' of resources was 'virtually impossible' with over nine hundred municipalities in Ontario. He added that the problem was compounded by the existence of too many special purpose bodies, which, despite past concessions to the OMEA, included local hydro commissions. In response, he announced that the government, rather than rely on local initiative, would take responsibility for reform to hasten its progress. The OMEA, Frame suggests, was not worried about the implications, believing reform to be McKeough's hobby horse rather than government policy. Indeed, Davis, in his part of the announcement, stated that he did not favour a 'single pattern' of regional government and that reform was not necessary throughout the province.[36]

Rationalization of local hydro was only the first of many disagreements between the task force and the OMEA. Both sought legal opinions during the spring of 1972 on the festering issue of who owned the HEPC, the government or the municipal hydro commissions. According to Dillon and Frame, legal recourse was a result of an accumulation of tension and frustration rather than a specific response to a particular event. The association was dismayed by the advance indications it had of the task force's recommendations. Besides having been invited for consultations, the OMEA had seen many of the task force's working documents, recalls secretary-manager Nokes, because Frame was a member of both bodies. From Dillon's perspective, the association rather than the HEPC was the real opposition to the task force. The difficulty was that its leaders honestly believed that the municipal hydros owned the commission and were incensed when the task force stated otherwise. He adds that because the HEPC had never contradicted the OMEA perception, the debate and the association's stance were from another age.

The legal opinion received by the task force was written by Blake, Cassels & Graydon. This firm determined that although the original and continuing contracts between the municipal hydros and the HEPC spoke of the latter as a 'trustee' for the former, the Power Commission Act, as a result of frequency standardization and consolidation of systems, had vested 'some' of its works and property in the HEPC 'absolutely.' Moreover, the opinion found that the sinking fund payments made by the municipal hydros, having never left the commission debt free, entitled them only to a 'beneficial interest in an unascertainable portion of the HEPC's assets.' It concluded that this municipal interest 'did not in any way relate to a right of ownership.' Nevertheless, to account for the payments, the opinion recommended that the municipal hydros be given non-voting participating share certificates.[37]

The opinion received by the OMEA was written, surprisingly, by Robert Macaulay, the former minister of energy resources and HEPC commissioner who was now a partner of Thomson & Rogers. Despite his past conflicts with the association and the commission, he was selected because of his expert knowledge of the legal issues. Ironically, Dillon, as executive director of the task force, also consulted with Macaulay many times because he was 'Mr Energy.' According to Frame, who familiarized himself thoroughly with the opinion, Macaulay's finding, like Blake, Cassels & Graydon's, was that the municipal hydros had an interest and the HEPC owned some property outright. Nokes agrees it was a 'hedged decision,' adding that Macaulay identified cooperative ownership as valid but difficult to prove in law. Although not able to recall the opinion in detail, Macaulay recollects that he told the OMEA it 'did not have a leg to stand on' on the ownership issue. The only exception was the question of whether the municipal hydros were entitled to compensation if the government decided to dismantle the sinking fund provisions of the Power Commission Act. In his view, these provisions were a Chinese puzzle created only as an illusion to honour the municipal contribution to the HEPC's origin rather than to confer ownership. While this was not what the OMEA wanted to hear, Frame recollects that Macaulay's opinion was of assistance because it also offered political advice on how to thwart the task force's expected recommendations for increased government control. Macaulay disagrees, believing that as a legal counsel he would only have enumerated his client's alternatives. Nokes accepts Macaulay's reading, and Frame insists that Macaulay's advice was definitely on the level of strategy and tactics.[38]

On yet another front of change, Davis gave recognition to public complaints over the HEPC's expansion. In June 1972 he appointed Omond Solandt, the chairman of the Science Council of Canada, as a commissioner to inquire into the route selection for a HEPC transmission line from Nanticoke on Lake Erie to Pickering east of Toronto. Born of property rights disputes, the Solandt Commission would primarily focus on matters of linear land use, but it would also make far-reaching recommendations on public consultation by the HEPC.[39]

Task Force Hydro's Report

The role and place of the HEPC was the subject of Task Force Hydro's first report, delivered on 15 August 1972. The general tenor of this report was that the HEPC should be re-established on terms comparable with other large private and public corporations and with more formal connections to the government. In this connection, its most significant recommendation was that the commission be designated a crown corporation and be renamed Ontario Hydro, as it had been known informally. Support for this recommendation, however, was not unanimous. In a written dissent, Frame argued that the term crown corporation implied government ownership and thereby was a derogation of the notion that the HEPC was the trustee of a municipal cooperative. He now adds that he then believed that the commission's exemption from the Crown Agency Act of 1959 would not sustain the recommendation, although he felt no obligation to bring this impediment to the attention of the other task force members.[40]

Frame's dissent was, to a degree, paradoxical because the decision to recommend a crown corporation structure was made to give the HEPC increased business autonomy from the government. In arriving at the recommendation, the task force had been struck by Herbert Morrison's description of a crown corporation in *Government and Parliament*: 'We seek to combine the principle of public accountability, of a consciousness on the part of the undertaking that it is working for the nation and not sectional interests, with the liveliness, initiative, and considerable degree of freedom of a quick moving and progressive business enterprise.'[41] Dillon adds that the term crown corporation was advocated to free the HEPC from political interference so that it could 'adopt the methods of a really slick commercial enterprise.' Nonetheless, Muncaster believes that the task force suggested the term, as the OMEA undoubtedly suspected, as a foundation recommendation which would 'resolve a whole lot of other questions,' the most contentious being ownership and control.

The creation of a crown corporation involved the replacement of the six-person commission with an eleven-member board of directors. Unlike the at-pleasure tenure of the HEPC commissioners, the task force specified tenure and categories for appointments: the chairman for a five-year term with unlimited renewal; two representatives of the OMEA selected from nominees of the association (which would formalize the practice begun by Premier Ferguson in 1925) for three-year terms twice renewable; five at-large directors, chosen for their expertise in matters related to the HEPC, with three-year terms twice renewable; and two senior civil servants as representatives of the government (in lieu of the informal practice of appointing government backbenchers) serving at pleasure. A president, appointed at the pleasure of the board and replacing the position of general manager, was to serve as an ex officio director. Believing that the presence of civil servants would make the HEPC 'much less responsive to the public than at present,' Frame dissented on this provision.[42] In hindsight, Dillon agrees that it was not a good recommendation but remembers it being offered to ensure that the board was constantly aware of the government's role.

In considering the mandate of the new corporation, the task force recommended that the HEPC be responsible for generation, transmission *and* distribution of electricity in the province in 'compliance' with the overall policy of the government. It was permitted, however, to delegate the distribution function to municipal hydro commissions. These recommendations were made in the knowledge that they trod on some of the statutory functions assigned to municipal hydros under the Power Commission Act and the Public Utilities Act but were offered on the grounds that the existing arrangements fragmented authority for decision-making on the total delivery system. In dissent, Frame argued that the HEPC's success was attributable in part to local control.[43] Reflecting on his dissent, he now adds that the recommendation was, in effect, advising the HEPC, as the child of the municipalities, to overtake the parents. This grandiose view of the municipal hydro role in the commission's operation, Muncaster recalls, made it very difficult to convey that the task force only wanted to take stock of historically assigned responsibilities.

Within the new mandate, the task force recommended that the chairman work full-time, with an outward orientation, translating, with the board of directors, government policy into 'corporate objectives and policies.' The president, for his part, was to be responsible to the board for the affairs of the corporation.[44] Given that the Power Commission Act had specified since 1955 that the chairman and the two vice-chairmen were the chief executive officers, the intent of this recommendation appeared to suggest that the president be the CEO. This would then have altered the division of labour between the executive committee or, more realistically given its inactivity, the chairman and the general manager. This division had historically been made along chief executive officer/chief operating officer lines.

Dillon recollects that the government had made it very clear to the task force that the chairman should not be the CEO, irrespective of past practice. Cronyn, for his part, goes further, arguing that there was no question that the president was intended to be the CEO with the chairman acting as a non-executive representative of the shareholders, meaning the government. Gordon agrees that the recommendation sought to place the actual running of the corporation with the president, but believes the intent was also for the chairman to retain a larger role than presiding at board meetings. Why the recommendation refrained from specifying that the president was the CEO, Dillon concedes, was that Gathercole 'would not have seen himself in any other role' but that of the CEO, and Davis was 'reluctant to upset [him] any more than he was already.' Gathercole, for his part, believes the recommendation was not specific because 'a good deal depended on the personality of the chairman and the president.'

By not recommending that either the chairman or the president be the CEO, the task force had not so much avoided controversy as decided that it was better not to make the designation. One of its members, Robert Taylor, leaning on his private sector corporate experience, had explained that designating the titles was less important than assigning the roles. This meant the division of labour between a chairman and president was better worked out between the two actors than superimposed, with room for flexibility to take account of personality, age, and workload. Taylor recalls stating that an ideal division of labour had the chairman and president metaphorically standing back-to-back, with the former looking out and the latter looking in. Since the task force was seeking a recommendation that would receive broad acceptance rather than cause unnecessary conflict, Taylor's view was adopted because it acknowledged the historic significance of a strong chairman who was good with policy and accommodated the desire for a professional in the presidency to run the corporation.

The proposed division of labour between the chairman and president was intended to enable the HEPC to respond to an increased volume of formal policy direction from government. With the chairman to provide the linkage, two interrelated recommendations of significance emerged. The first was that the cabinet 'give expression' to the corporation's 'broad objectives and constraints' through a contract, an idea borrowed from Electricité de France.[45] While this general parameter was advocated to enhance operational autonomy and reduce the need for ministerial supervision, the second recommendation was that the provincial secretary for resources development set government policy not specified in the contract, but only after consulting the board of directors. This would make the policy minister the commission's point of contact with the government, even though these ministers were not intended to have line responsibilities. According to Cronyn, the two recommendations were offered in lieu of making the HEPC responsible to the minister of the environment. The purpose was to create at least the appearance of a more distant relationship, one where the provincial secretary would only answer for policy and not programs.

More generally, the task force recommended that the HEPC was to participate actively in the fulfilment of government energy and environmental policy, with environmental costs included in its rates, and continue close coordination with the government on financial matters. However, it did recommend that the commission charge the government for the cost of supporting 'regional development or counter-cyclical' policies rather than build their costs into rates.[46] According to Muncaster, the latter was offered, in part, in response to the rate deferral of 1972 but more generally as a signal that consumers should not pay for, in the jargon of the day, 'social engineering.'

As for the question of who owned the HEPC, the government or the municipal hydro commissions, the task force played down the issue by sticking to the legal advice received from Blake, Cassels & Graydon. While it recommended that control and ownership 'continue to reside' with the government, the municipal sinking fund contributions were to be accounted for as equity – a word of some significance – through non-voting participating shares redeemable only if the corporation was wound up.[47] In making this concession, Dillon recollects that the task force knew that the OMEA leaders would 'blow their stacks' if the ownership issue were resolved definitively in the government's favour. Given this parameter, Dillon recalls that the task force had decided to make the question of who makes policy the more important issue. Indeed, Taylor remembers agreeing with the share scheme to appease the OMEA, even though he felt the municipal cooperative contention was nonsense. His reason was that the shares would be of little or no value because ownership without control meant nothing, and because the HEPC was a closed system where equity could not be exchanged, let alone command dividends.

The recommendations on ownership, despite their recognition of a municipal equity interest, nonetheless diminished the OMEA's longstanding claim that the HEPC was the trustee of a municipal cooperative. Surprisingly, Frame did not offer dissent, although this was one of a number of issues where he was torn. He recalls that he felt it was impractical for the OMEA to take a normal ownership role and, short of doing so, the share certificate scheme gave 'significant recognition' to the financial interest of the municipal hydros. For this stance, which the OMEA itself refused to accept, Dillon remarks that Frame was 'no messenger boy' for the association.

Where Frame dissented the most vigorously was on the recommendations respecting regional government, an issue which had consumed the OMEA since the Beckett report in 1964. Although the association had thought that the battle for flexible implementation had been won, the task force recommended that local hydro should be organized uniformly as an upper-tier municipal function for economic and planning reasons. It also advocated administration by the chairman and appointees of the regional council in the belief that it was too difficult to elect commissioners at the upper tier and inappropriate for hydro to be run by an elected council. Drawing attention to the lack of consideration for local costs and geography, as well as the tradition of elected commissions, Frame claimed that these recommendations would not improve service and would likely result in higher rates.[48]

Before releasing the role and place report, Davis asked the OMEA for its views. In its 19 September 1972 response, the association reiterated Frame's points and added its own objections, especially to the ownership and control recommendations. Sticking to its municipal cooperative contention, the OMEA's argument was that power at cost, the HEPC's ostensible operating philosophy, had meaning because it was in essence a scheme of 'cooperative accounting.' The OMEA closed by placing itself squarely with the status quo:

> The province will be better served by an organization that has its roots at the consumer level with proven experience and ability to meet the needs of the consumer and cope with required change. This is not the time to switch to a proposition that offers arbitrarily increased costs, no improvement in service and a more autocratic and remote corporate structure.[49]

Thus, the association was not persuaded that the role of municipal hydro would not change a great deal as a result of the task force's recommendations.

Despite the OMEA's reaction, Davis endorsed the crown corporation recommendation when the report was released publicly on 13 November. He added that the new corporation would be responsible to a minister without portfolio who was a member of the resources development policy committee, rather than the policy minister as had been recommended. He did, however, reject the recommendation for civil servants to sit on the corporation's board.[50] Although Dillon recalls that the OMEA's opposition had been a contributing factor, Davis recollects doing so because it offended his own sensibilities.

The Solandt Commission delivered an interim report on the transmission line it was studying on 31 October 1972 calling for an 'open planning process.' Although Solandt found no real evidence to suggest the HEPC wanted to keep its plans secret, he did find that it had been uncommunicative with the public. Rather than subject its business autonomy to stricter government control, he concluded that, on its own, the commission should expose its plans at an early stage to public scrutiny. Task Force Hydro's second report of 14 December would concur with Solandt. As for the larger environmental trade-offs of HEPC projects, Solandt tentatively felt the government might need to create an independent advisory council or a body with quasi-judicial powers.[51]

Fulfilling another dimension of the review process, the Advisory Committee on Energy reported on 19 December, expressing the ominous warning that 'the period of seemingly unlimited abundance and cheap energy has come to a close.' In its view, there was an immediate need for long-range policy formulation and planning capability in the Ontario government, whose existing capability it considered to be largely uncoordinated and entirely inadequate. In a general dissent, however, Henry Regier, the Pollution Probe representative, objected to the advisory committee's unquestioned assumption that energy demand would continue to increase. By refusing to consider alternate demand scenarios, his concern was that this view might stimulate growth, and thereby become self-fulfilling.[52]

The committee's main recommendation was that the government should establish a ministry of energy, whose purpose it would be to develop 'comprehensive and coordinated' energy policies. Such a ministry would facilitate a 'consistent policy approach' and provide for a 'clearly designated' minister responsible to the legislature, purposely placing the HEPC and the Ontario Energy Board under the same minister.[53] Thus, despite the joint membership of Dillon and Reynolds, the committee went in the opposite direction to the task force, which had recommended that the commission be connected to the government through the policy minister. While their different mandates and memberships are the explanation, Dillon feels that the task force was distinguished by an anti-bureaucratic attitude. As for Hay of the OMEA and Gathercole, both of whom were advisory committee members, neither registered dissent given that the HEPC had already been made formally responsible to the minister of the environment. Gathercole, for his part, remarks that he had become resigned to increased government supervision of the HEPC.

Until a new energy ministry could be established, the advisory committee called for the creation of an Ontario Energy Commission. It was to report, in the interim, to the provincial secretary for resources development, assuming the functions proposed for the new ministry and other permanent functions. The latter were the coordination of environmental assessments for proposed energy projects, both private and public, the regulatory function of the Ontario Energy Board with respect to the natural gas industry, and the power to review, although not regulate, the HEPC's rates, long-term plans, capital spending requirements, and energy choices, including nuclear power.[54] In arriving at its decision to recommend review rather than regulation, Dillon recalls that the committee was seeking to get decision-making on rates 'out of the premier's office.' Thus, the recommendation for review was also an indictment of the inadequacy of the OMEA's oversight, which, Reynolds notes, was in conflict of interest as the principal electric lobby. Moreover, it conveyed the irony that the OMEA had not been a major player in rate determination.

Hay viewed the recommendation for the Ontario Energy Commission as authority for full regulation. In dissent, he claimed review would either become perfunctory or an intolerable bottleneck. Without an accepted measure of environmental cost, the recommendation for assessments could be read as favouring only risk-free developments. As an ardent non-interventionist, he argued that with one government agency scrutinizing another, there would be 'bureaucratic delay, procedural complexity, diffusion of responsibility and duplication of oversight and advice to the Cabinet.'[55] Gathercole did not join Hay in dissent, although he had opposed regulation at the 1972 OMEA convention. He recalls accepting review because it was a compromise, one which Gordon remembers would be helpful to the HEPC and the public and take the heat off the government.

The Municipal Campaign against Reform

In the aftermath of Davis's acceptance of the task force's crown corporation recommendation, the OMEA undertook a campaign to block its implementation. Its executive, with Frame in support, encouraged member municipalities to insert a flyer in their retail billings beginning in early January 1973. The flyer luridly contended that the task force sought to subsume local hydro into 'a huge, single administrative system centred at Queen's Park'; that the system would be directed 'mainly by politicians and civil servants' (which was ironic because the OMEA leaders were themselves politicians); and that the purpose was to institute a 'vast, profit making commercial enterprise' in place of power at cost, with the government able to 'incorporate "hidden taxes" into the rate structure.' The flyer closed by encouraging consumers to write to the premier and MPPs to 'oppose this "take-over" of *your* electricity system.' Given broad distribution in a time before billing inserts were common, the flyer was a success, bringing an onslaught of public inquiries and protests.[56] Davis, on receiving a flyer at his own home, recalls finding humour in being encouraged to protest against himself, but concedes that the OMEA campaign was disconcerting because the government had traditionally had good relations with the association. Dillon feels that the OMEA honestly believed the statements in the flyer and had not intended to mislead the public. But he remembers, indicating the task force's frustration, that the OMEA was dubbed the 'electric shriners' for its antics.

The fall-out from the flyer was pronounced. Having personally received a series of blasts from task force members and politicians, Frame suggests it would be an understatement to say the government was disturbed. To enable cabinet ministers and MPPs to respond to the public reaction, the task force issued an information statement on 10 January, labelling the flyer's criticism as negative, misleading, and erroneous. The task force considered the flyer particularly galling because it had kept the OMEA informed of its work and had considered the two to be in essential agreement. Davis responded directly on 15 January by appointing McKeough, who had resigned as treasurer in September 1972 over a conflict of interest while minister of municipal affairs, as his parliamentary assistant with special responsibility for energy.[57] Although McKeough's specific mandate was to report on how the recommendations of the task force and the advisory committee could be 'integrated and expedited,' Davis recollects that part of his responsibilities was also to handle the OMEA's opposition.

With this political wrangling in the background, the Ferguson formula for appointments took a drubbing when A.A. Kennedy was encouraged to resign from the HEPC on 31 January. He had served for eighteen years since first being appointed in 1955 while president of the OMEA. Muncaster was immediately appointed in his stead.[58] Having had no notice of this departure, the OMEA was unable to request that its representation be renewed, let alone suggest a new nominee. The reason, Dillon maintains, was that the government had chosen to limit the OMEA to two representatives as the task force had recommended. However, he had discovered while observing HEPC meetings, much to the chagrin of Gathercole, that there was a notorious lack of policy discussion. After Gathercole would explain a matter at hand, Flemming, one of the association representatives, would inhibit discussion with the standard response: 'Well George, I think you have said it all.' For this reason, Dillon concedes that Muncaster, with his corporate experience, was appointed to give the HEPC more 'competence,' and to ensure that there would be no 'drifting' before the task force's recommendations could be implemented. The OMEA, Frame admits, was less concerned about its diminished influence than with the prospect that it might not be included on the new Ontario Hydro board of directors.

The business of Task Force Hydro proceeded apace, with its third report, on nuclear power, delivered on 16 February. Accepting as given that growth in demand would continue and that the HEPC would need to increase capacity, in part, with nuclear power, it made three groups of recommendation, mapping an outline for the commission's anticipated needs to the year 2000. First, the task force called on the commission to standardize its use of CANDU technology, assure itself supplies of uranium through long-term contracts, and ensure supplies of heavy water by developing its own if necessary. Secondly, the task force called on the HEPC to support and promote diffusion of the technology abroad in cooperation with AECL to reduce overall costs. And finally, it called for the HEPC to promote education and information. The last was to include a major and sustained campaign to improve the public's knowledge of nuclear technology and enhance its appreciation of the importance of the effective exploitation of nuclear energy for the economy of Ontario.[59]

By making these recommendations on nuclear power, the task force was also making a statement on expansion and on the economic impact of energy choices. On the former, it worked from a starting point of the HEPC's own forecasts on demand. Although in an earlier era great credence had been placed on the historical rate of growth of 7 per cent per annum, the forecasters were looking ahead but still seeing the past. As for the latter, there were at least two reasons for going nuclear. Dillon recalls that its economic impact for Ontario was an important consideration, while Reynolds believes the dearth of other generation sources in Ontario was the motivation. Gordon maintains that both views had prevailed, and Taylor links them together with the economic nationalist view that expenditures in Canada over the life of a nuclear plant were greater than that for a fossil fuel plant.

The task force's first report dominated discussion at the 1973 OMEA convention. As an indication of the association's still defiant mood, President A. McGugan stated that while some recommendations were 'changes for the better,' others would 'serve the public no better and probably not as well.' He was careful not to lay blame on Frame, whom he felt had championed the municipal hydro cause but been badly outnumbered. He was nonetheless optimistic that a 'new, vigorous and harmonious' relationship between the 'electrical system' and the government would result.[60] The contentious issues were addressed before the delegates in succession by Davis, Gathercole, and Muncaster on 6 March.

In only the fourth speech ever by a premier to the association and the first since 1963, Davis used some levity to break the tension in anticipation of his remarks. Much to the OMEA's dismay, he defended the task force's ownership and control recommendations. Indeed, he specifically stated that the HEPC was an agency of the government and not a cooperative of the municipal hydro commissions, although he accepted the share certificate scheme. As for the crown corporation recommendation, which he had previously accepted, he stated that it was intended only to give the HEPC a 'structure to suit contemporary requirements,' adding that the task force had not recommended a take-over of the municipal hydros or their financial stake in the commission (he refrained from calling it equity). Moreover, he rejected the argument that profits from the new corporation would be transferred to the government, stating that electricity would continue to be supplied at cost. As for the task force's expected recommendations, Davis revealed that he favoured rate review, although not regulation, and possibly an environmental review board, as Solandt and the advisory committee had recommended.[61]

Despite his support for the ownership and control recommendations, Davis did make some concessions to the OMEA's view. He repeated his objection to the recommendation that civil servants be appointed to the board of the proposed corporation. He also rejected the recommendation that local hydro should uniformly be a regional government function, favouring, instead, a case-by-case determination. Further, he supported the continued election of hydro commissioners, as long as there was statutory flexibility to appoint them where appropriate. While Dillon considers the regional government concessions as a trade-off for the ownership and control recommendations, Davis does not agree that he was seeking to soften one set of recommendations by rejecting the others. In assessing what would be acceptable and workable out of what the task force thought to be the right structure, he recollects that he simply 'had not settled in [his] own mind that regionalization was necessarily the best way to go.' Indeed, he had stated as much in Design for Development in 1972.[62]

Whether or not Davis had sought to appease the OMEA, his concessions had little effect. He was 'torn to pieces,' much to the astonishment of the head-table guests, in the customary thank you address given by Gordon Robertson, an OMEA board member.[63] Davis's recollection is that Robertson had been the appointed spokesperson for the association, presenting prepared remarks without 'totally agreeing with what he was saying.' Frame takes a different view, explaining the mishap as a result of poor communication, with Robertson not having been informed that Davis would offer concessions. Although only Gathercole believes Robertson was acting as a partisan, Nokes, the secretary-manager, reveals that Robertson was an ardent Liberal, and Frame admits that he was a good friend of Liberal leader Robert Nixon. In McKeough's eyes, the whole affair of the OMEA's objections to reform was troubling because Davis felt an affinity with the association as a former HEPC commissioner.

When Gathercole took the podium following Robertson's comments, he encouraged the delegates to recognize that the task force's recommendations were a 'serious and genuine effort to strengthen Hydro.' In the remainder of his speech, he addressed the public relations challenge that the HEPC and, by association, the municipal hydros faced. He stated that despite the commission's best efforts to the contrary, it was perceived as 'remote, impersonal, monolithic and indifferent.' His hope was that making a commitment to open planning, as Solandt and the task force had recommended, would improve matters, but he recognized that the efforts could very well be self-defeating.[64]

The delegates, unappeased by Gathercole's speech, were seething when Muncaster took the podium to take questions from the floor.[65] Among the generally hostile questions he was asked, the most noteworthy dealt with the HEPC's ownership and the OMEA's representation on the proposed board of directors. On the former, the designation of the HEPC as a crown corporation was objected to because it denoted ownership by government when the municipal hydros paid the commission's debts. Muncaster responded that the true test of ownership was who calls the shot, which left him no doubt that ownership was vested with the government. He added, bluntly, that the municipal hydros had only paid for some of the debts, not all, to which Robert Taylor, who accompanied Muncaster on the dias, added that the task force had recommended that the municipal stake in the HEPC be made more explicit through the share scheme. Similarly, on the OMEA being limited to just two of the eleven directors, Muncaster responded that directors serve the corporation's interests rather than their own. He added that it would be up to the association's representatives to convince the other directors of the municipal case.[66]

Reflecting on the OMEA's actions, Reynolds believes it was fighting a desperate battle to retain influence. Frame, on the other hand, believes that the OMEA's resistance to change arose because the task force's ownership and control recommendations were not well explained. As for the assault on Muncaster, while Taylor recollects that it reflected the views of the majority, Frame remembers it as coming from a vocal minority. Although many delegates were familiar with modern corporate organization, Frame feels that those from the small municipalities and those with small business backgrounds viewed Muncaster as a 'high corporate financier.' Muncaster agrees that the moderates, however numerous, were not vocal, but he contends that a large number of delegates, including many of the executive, felt they were being sold out. In any event, a majority of delegates sought to reverse the tide towards the creation of a crown corporation in the resolutions session which followed. The OMEA again endorsed the principle of a capital levy as a means to decrease government influence and called for half of the proposed corporation's directors to be appointed from OMEA nominees.[67]

Following the convention, the OMEA's distaste for the term crown corporation received support in an unlikely turn of events. The COGP's belated agency recommendations, delivered in March 1973, held that all commercial ventures of government simply be called corporations. Since the task force was the creation of the COGP, this effectively vetoed the former's recommendation that the HEPC be designated a crown corporation.[68]

After receiving the advice of Task Force Hydro, the Solandt Commission, the Advisory Committee on Energy, and the COGP, the government announced in the speech from the throne on 20 March 1973 that 'new policy initiatives establishing agencies for energy may be expected as well as changes in the role and place of the HEPC.' To facilitate this end, McKeough, as parliamentary assistant to the premier, submitted an interim report to Davis on 2 April, outlining 'an administrative and reporting structure' for energy matters in the government of Ontario. His main recommendations followed those of the task force rather than of the advisory committee. Although Muncaster remembers being surprised by this outcome, notable among those who advised McKeough were Dillon, still executive director of the task force, and Robert Macaulay, who had been dismayed by his experience as minister of energy resources from 1959 to 1963.[69]

McKeough recommended that an energy secretariat be established under the provincial secretary for resources development. The secretariat was thought to be preferable to the advisory committee's call for an energy ministry because it would elevate energy policy to an 'overall' level for the line ministers on the resources development policy committee of cabinet and because no operating programs were required. McKeough recalls that he also did not follow the advisory committee's model because he was rumoured to become minister of energy at this time and did not want to appear self-serving. In addition, the HEPC was to be made responsible to the provincial secretary rather than to a line minister in the policy field so as not to leave potential for conflict between ministers. Recognizing that he was advocating a quasi-line responsibility for a policy minister, McKeough cleared this breach with COGP chairman John Cronyn on the basis that it was better than creating a new ministry and would give the policy minister, whose function was not clear, added purpose. McKeough also recommended that the Ontario Energy Board's regulatory role for the natural gas industry be expanded to include review of the HEPC's rates, with the board also to be made responsible to the provincial secretary.[70] His reason, he recalls, was that 'public outcry' demanded that there at least be 'outside review.'

While McKeough's report was under consideration, Task Force Hydro delivered its fourth report, on financial policy and rates, on 11 April. Although it found that no useful purpose would be served by abandoning the government guarantee of the HEPC's debt, its other recommendations involved a dramatic departure. They posed many challenges to the provisions of the Power Commission Act that sustained the OMEA's contention that the HEPC was the trustee of a municipal cooperative. Muncaster, Reynolds, Dillon, Taylor, Gordon, and Frame all concede that the task force knew that the recommendations would undermine this contention, although none recalls the report being held over until after the OMEA convention to avoid controversy.[71] Since the task force had discussed the cooperative issue at length, and done so with Frame an active participant, Reynolds recalls the outcome as realistic rather than part of a deliberate strategy to undermine the OMEA. According to Taylor, the provisions simply impeded regularizing HEPC practices with the private sector. For this reason Gordon, from the HEPC's perspective, admits that they were passé.

The most significant challenge to the municipal cooperative contention affected the sinking fund payments by the municipal hydro commissions. As a source of internally generated funds, the task force found this debt retirement scheme unorthodox because, as Muncaster explains, it produced double depreciation. Moreover, the payments, restricted by statute to a fixed percentage, had failed to generate an adequate rate of return. As a result, the HEPC had operated with a high debt/equity ratio, a policy which the task force felt should not be continued. Instead, it recommended a non-statutory municipal charge set by the board of directors for both debt retirement and system expansion. The objective was to give the HEPC the flexibility to determine the mix of internally and externally acquired funds which minimized the overall cost of capital. With the municipal cooperative contention in jeopardy, Frame dissented on this recommendation.[72]

On the matter of rates, the task force made two recommendations that would directly undermine the OMEA's cooperative contention. The first was to discontinue the practice of charging an interim rate for twelve months with a thirteenth bill for adjustments. Although this scheme was acknowledged as an important symbol of provincial and municipal hydro commission cooperation, the task force reasoned that the recent major adjustments had come from the rate stabilization reserve. The thirteenth bill was said to be minor in comparison and set at the discretion of the HEPC. From the municipal perspective, Frame felt the scheme was the foundation of a cooperative enterprise, and for this reason he dissented.[73] Muncaster concedes that the scheme was very important psychologically to the municipal hydro leaders but believes it was nonetheless an anachronism.

Following the lead of the advisory committee and McKeough, the second recommendation of significance on rates was to establish external review. The task force had in mind a 'broadly based public forum and rate appeal mechanism,' and Frame did not disagree that the internal check and balance of the OMEA's power-costing committee was insufficient. Formal regulation was rejected because it would duplicate the function and expertise of the proposed board of directors.[74] With public openness the main objective, Dillon recollects that the task force was 'absolutely sold' on the opinion that review was better than regulation. Review placed the onus on the HEPC to make its case for an increase, rather than on the regulator to provide it with a sufficient increase. This avoided the regulator becoming sympathetic with the regulated and forced the HEPC to assume full responsibility if it disagreed with the review body. Muncaster adds that the task force favoured review to ensure 'regulatory lag' did not hold up timely rate increases. The HEPC, not surprisingly, favoured having this final authority on rates, because it was the one running the business.

In conjunction with bringing rate-making out into the open, the task force made other recommendations on principles to guide decision-making. It stated that there were 'inherent dangers' when a crown corporation with the commercial mandate of the HEPC 'attempts to deal directly with economic and social maladjustment.' Although Dillon recalls that this position was offered in response to the 1972 rate deferral, other task force members do not make this direct connection. In general, the view was that rates should reflect the costs of production,[75] which Muncaster recollects meaning that the government should be responsible for costs beyond the basic mandate of the HEPC.

Before the government could move on the statutory front it was rocked by a political scandal over the HEPC, dubbed by the media as 'Hydro-gate,' in late April 1973. At issue was a lease-purchase contract the commission had signed in October 1972 for its new head office building, Hydro Place, then under construction. The builder, Canada Square Corporation, was controlled by Gerhard Moog, a personal friend of Davis. With rumours of impropriety circulating and journalists onto the story, on 27 April Davis tabled in the legislature a letter from Gathercole explaining the process that led to the selection of Moog's company. It stated that the HEPC had sought and evaluated four competitive proposals before deciding that the 'design proposed by Canada Square was superior to the design discussed and proposed by the others.' However, on 28 April the *Globe and Mail*, in a headline story, revealed that Moog's proposal was in fact based on architectural plans for which the HEPC itself had paid nearly $1.5 million in 1970 prior to cancelling plans to own the building outright.[76]

This disclosure was followed by another front-page story on 30 April which further contradicted Gathercole's contentions. It revealed that the HEPC had made no public announcement of a competition, that the three developers other than Moog had had to approach the commission themselves, and that the information given to them, in contrast to Moog, was 'so general as to be useless.' Moreover, one of the developers, later revealed to be prominent Liberal Don Smith of Ellis-Don Limited, alleged that a person close to the cabinet and high in the Conservative party – a composite for COGP chairman John Cronyn and realtor J.J. Barnicke – had told him 'to keep [his] mouth shut or [he] would never get another government job.' To make matters worse, Gathercole was quoted as still maintaining that all four companies had been given an equal chance to receive the contract.[77]

With Gathercole's integrity having been questioned and his own credibility at stake, Davis announced that afternoon that a select committee would investigate the Hydro Place contract. Established on 1 May with broad terms of reference under John MacBeth, a government backbencher who had been an OMEA vice-president when first elected in 1971, it would not report until October after the legislation to revamp the HEPC had been passed.[78] Although Davis does not recall the affair or the committee's proceedings affecting the content or timing of the legislation one way or the other, his view was not shared by all. James Fleck, the COGP executive director who had succeeded Reynolds as deputy minister to the premier, concedes that the affair was 'very distracting.' McKeough, who was charged with drafting the legislation, believes the incident actually spurred the government into action by making increased control over the HEPC a 'good thing to do politically.'

In the shadow of the inquiry, Gathercole responded to all of the reviews of the HEPC in its 1972 annual report, delivered 25 May 1973. On Task Force Hydro, he stated that a number of organizational changes were forthcoming. On the Solandt Commission, he revealed that HEPC would commit itself to open planning, inviting property owners, environmentalists, ecologists, and conservationists into discussions on alternative development plans. On the Advisory Committee on Energy's dire warning of energy shortages, he remarked that all indications suggested that electricity would have an expanded role in meeting Ontario's energy requirements as well as its economic health and environmental quality. More specifically, he stated that the warning underscored how the HEPC's nuclear program would play a vital role in Ontario's energy future. He predicted that nuclear power, then only producing 9 per cent of the HEPC's output, would produce 60–70 per cent by 1990.[79]

The HEPC's expansion plan based on nuclear power was of great significance. By all accounts, it was an industrial development program, although not one explicitly based on a government policy of energy security. Gathercole actually goes further than others, calling it an industrial strategy and Ontario's mega-project. Although the government was not in the lead, Macdonald, then deputy treasurer, concedes that a common understanding on the economic impact of nuclear power was developing 'largely, as much as anything, on the advice and impact of Gathercole.' Indeed, expansion came to be part of a broader government economic strategy, one where decisions were made around the HEPC's capital spending. Given this degree of informality, Dillon remarks that the expansion plan was conceived in an uncomplicated age by a few people who saw what the future held, adding that this could not have transpired since without full public scrutiny.

In anticipation of an energy crisis, on 7 June 1973 Davis provided an extensive outline of new government policy on energy matters in the legislature. He began by announcing the government's approval in principle of a $3.5 billion HEPC capital program. Its main elements were a second Bruce nuclear plant, the purchase of AECL's heavy water plant at Bruce and the construction of a second, and interim approval for a new nuclear plant at Darlington. The program also included, ironically, an oil-fired plant at Wesleyville, later mothballed because of high oil prices. On the thorny issue of rationalizing local distribution, Davis announced that a committee would develop principles and guidelines for a case-by-case examination, which was consistent with his commitment to the OMEA. And following through on plans to revamp the HEPC's character and its relationship with the government, Davis personally introduced the legislation.[80]

The Power Corporation Act

Through a matrix of three statutes proclaimed on 22 June 1973, the HEPC's character was modernized and its relationship with the government was given a new, more formal complexion. Broadly speaking, the Power Commission Amendment Act reconstituted the HEPC as Ontario Hydro with a corporate structure; the Ministry of Energy Act established a new portfolio; and the Ontario Energy Board Amendment Act established external review of Ontario Hydro.[81] When compared, however, to the volume of advice received from the various inquiries, the alterations to the

HEPC and the revisions to its relationship with the government were relatively few in number.

In explaining the paucity of amendments, Robert Taylor, who would become chairman of Ontario Hydro, feels that the government instituted the basic structure for the corporation, the ministry, and external review and left the remaining recommendations to the discretion of the new actors. Davis concurs, because a government, faced with a massive number of recommendations, would tend to act on what it could digest and leave the rest to evolve. He believes his government implemented the basic recommendations, believing the others that made sense would flow from them. In choosing this route, Reynolds suggests Davis's speech to the OMEA had been a trial balloon for the statutory changes, and Dillon adds that the alterations were not extensive because Davis had not been received with 'deafening cheers.' Many of the recommendations, as Cronyn and Gordon recall, involved operational matters requiring action from the new board of directors rather than the government. Nevertheless, Fleck, from the premier's office, remembers that the government was reluctant to make too many matters rigid through statutory enactments.

The amendments to the Power Commission Act, including its new name, the Power Corporation Act, were limited to replacing the commission structure with a corporation structure. In essence, the provisions for a six-person commission, its executive committee of three chief executive officers, and the dormant Hydro-Electric Advisory Council (created in 1944 but never made operational) were repealed. In their stead, a thirteen-person board of directors was established. This would consist of a chairman and up to eleven other directors appointed by the cabinet, a vice-chairman selected by the board from among the eleven, and a president as an ex officio director. The chairman was to serve for a five-year term with unlimited renewal and the directors for three-year terms, twice renewable. The president, installed by the board, was to serve at pleasure. In place of the leave required from the attorney general to launch suit against the HEPC or its commissioners, the new board and directors received a general indemnification.[82] In the process of creating a corporate structure, the legislation enhanced the arm's-length relationship between the government and Ontario Hydro by strengthening the position of the board as a buffer.

Attendant to the corporate focus, there were two other amendments of note. The first was that the act now directed Ontario Hydro's annual report to the new minister of energy rather than the minister of the environment.[83] In adopting this recommendation of the advisory committee, the government rejected the task force's view that the HEPC have a direct relationship with the provincial secretary for resources development. According to McKeough, who re-entered the cabinet at this time as minister of energy, the policy minister route was not considered a sufficient response to the changed energy environment, although he also recalls that Davis was not enamoured of the role of these ministers. Reynolds, who was then deputy provincial secretary for resources development, feels that Davis had never really been committed to the concept of policy ministers, and Muncaster agrees that their role was already falling out of favour.

The second amendment was that the chairman was designated the chief officer, although not chief executive officer, and was to continue to be full-time.[84] Although this designation had not been made clear in the task force's original recommendation, it was included, Cronyn recalls, as a result of Gathercole's continued insistence that the chairman should be the CEO. According to Dillon and Muncaster, the term chief officer was a compromise arrived at among McKeough's circle of advisers. It was adopted with Davis's consent, McKeough remembers, in the knowledge that the term occurred nowhere else, and nobody knew what it meant, except that the president was going to be more important and the chairman was going to be the public face. Gathercole, for his part, believes that despite the chairman's designation as chief officer, the president was to be the CEO.

With the balance of the old Power Commission Act left intact, the task force recommendations most notable by their absence were those that would have undermined the notion that the HEPC was the trustee of a municipal cooperative. Of greatest significance, Ontario Hydro was not termed a crown corporation, the share certificate scheme was not implemented, and the statutory provisions for municipal hydro sinking fund payments and the thirteenth bill remained intact. Moreover, the HEPC's exemption in the Crown Agency Act of 1959, of which only the OMEA seemed to be aware, continued to apply to Ontario Hydro.[85]

Frame views this general outcome as a victory for the OMEA's campaign and evidence that Davis recognized there was no political gain and plenty of political flak if he acted on the task force's recommendations. Davis concedes that the term crown corporation might have been avoided deliberately, but believes that the government was less dissuaded from acting on the recommendations in question than it was persuaded that not much would be different if more of them had been implemented. Nevertheless, Dillon maintains that OMEA opposition had been a positive factor in the government's retreat from the ownership recommendations and that Davis was reluctant to challenge the OMEA any more than necessary.

There were only two other recommendations of significance notable by their absence in the new Power Corporation Act. No provision was made for a contract between the government and the HEPC outlining the latter's objectives and constraints. Having been offered in tandem with the policy minister being made the responsible minster, it was left in limbo. Although both Cronyn and Fleck suggest that the contract issue was left non-statutory for purposes of flexibility, no contract would appear without statutory instruction. There was also no amendment relating to regional government, given Davis's expressed view on the matter.

Although the Power Corporation Act gave the minister of energy the single function of receiving Ontario Hydro's annual report, the Ministry of Energy Act gave the minister a larger, although ambiguous role in Ontario Hydro's affairs. The minister was given responsibility for the administration of the Power Corporation Act and the Ontario Energy Board Act. Moreover, the stated functions for the minister and objectives for the ministry were to review Ontario's long- and short-term energy needs on a continuing basis; advise and assist the government on intergovernmental energy matters; and make policy recommendations, especially with regard to supply, prices, development of indigenous resources, and research matters, including conservation and efficiency.[86] According to McKeough, Davis chose to follow the advisory committee's recommendations and create the new ministry both to elevate energy issues politically and to create a portfolio where he could rehabilitate himself.

The broad consequence of establishing the Ministry of Energy, according to Reynolds, was that it emasculated the role of the provincial secretary for resources development in energy policy. While Dillon concedes that there was no question that McKeough was responsible for energy policy while minister, he nonetheless argues, having been the first deputy minister of energy and then succeeded Reynolds as deputy provincial secretary, that the result depended on the actors. McKeough, for his part, remembers that the energy portfolio had a negative effect on the policy minister, although he adds that he himself had little to do besides 'talk about wind power and conservation' unless he 'grossly interfered' with Ontario Hydro or the Ontario Energy Board. Indeed, Muncaster, who was an Ontario Hydro director until 1980, recalls that the corporation dominated if not overwhelmed energy ministers by its large presence in the policy field. In his view, the government might have been better off if Ontario Hydro had been linked to the policy minister, as was recommended, thereby releasing itself from responsibility for programs. Davis disagrees with this conclusion on the grounds that all ministers were equal and that the principle of ministerial responsibility would not have varied. In any event, Douglas Gordon, who would become president of Ontario Hydro, remarks that a more direct linkage with the government resulted than the task force had thought appropriate.

The amendments to the Ontario Energy Board Act centred on giving the board, which already regulated the oil and natural gas industries, the power to review, although not regulate, Ontario Hydro's rates and business affairs. The board, which had been under the purview of the energy resources portfolio beginning in 1960 before being moved to mines and northern development in 1970, was also made responsible to the minister of energy. This was done for functional reasons, with no acknowledgment of the conflict presented by the minister being responsible for the operating agency and the review agency. To facilitate the review of Ontario Hydro, the existing general provision of theOntario Energy Board Act, which allowed the cabinet to refer for review any energy matter, acquired two new provisions. The first affected electricity rate increases as from 1 January 1975. Hydro was to submit rate proposals for municipal and direct industrial customers to the minister eight months before the rates were to take effect, with direct rural customers unaffected. The minister was then to refer the proposals to the energy board, which in turn was to hold public hearings and make a report four months before the rates were to take effect. However, the board was not given the quasi-judicial power to set rates, and the minister was not given the power to arbitrate an impasse between the board and Ontario Hydro, despite being responsible for both.[87] As a result, Hydro continued to have final authority to set rates for all classes of customers.

The arrangements for review rather than regulation were agreed upon, McKeough recalls, as a compromise that made the OMEA and the HEPC happy, although he himself never favoured regulation. According to Muncaster, both regulation and ministerial arbitration were rejected because the government decided that the ongoing viability of Ontario Hydro should be left with the board of directors. Taylor adds that if the minister, without the expertise of Hydro or the energy board, were given the power to arbitrate in either's favour, the decisions would have been suspect. Moreover, with the expertise elsewhere, he feels the minister did not need to develop the capacity for another review. From the utility's perspective, Gordon remarks that it was possible the government did not want the power to arbitrate rate disputes, and Frame concurs that the government had wanted to 'isolate itself from the rate situation.' In this vein, Fleck remarks that the objective was only to make the rate process transparent.

By enacting only review, the government had left open the possibility of introducing full regulation at a later date if review proved unsatisfactory. Indeed, Macaulay recalls looking at review as only an interim arrangement, while Dillon remembers hoping this would not be the case. On this score, Dillon remarks that the absence of regulation did not diminish the power of 'letting people have their day in court.' Macaulay, who would serve as chairman of the Ontario Energy Board from 1984 to 1987, counters that public and media scrutiny 'has not been worth a damn' because it has been incapable of grasping the complicated and technical nature of the issues. Whatever the case, Davis does not recall there ever being a plan for regulation because there was not the same need as there would have been with a series of private utilities and because Ontario Hydro was non-profit. Moreover, he recollects, very tellingly, that rates would, in fact, be approved by the cabinet, although he does not recall a case where the government imposed rates. Despite the lack of a statutory provision for this practice, it existed, he concedes, because 'the government ultimately had to accept responsibility.'

The second new provision under the Ontario Energy Board Act was that the minister of energy was given the power to refer to the board other Ontario Hydro matters. This provision specifically included the 'principles and practices of power costing, rate making, financing, service reliability, system expansion and operations.'[88] This meant that the cost factors underlying rate increases were to be reviewed only at the minister's discretion. The separation of costs from rates arose, McKeough recalls, because he wanted to avoid the situation of natural gas regulation, where all cost issues were examined for every rate increase. He had in mind that the board would be referred contentious cost issues, hopefully covering the full gamut of Hydro issues over a cycle of approximately five years. He adds, moreover, that although the cost issues were to be referred at the minister's discretion, the energy board had the prerogative to make requests for referrals.

The most striking feature of the new Ontario Hydro board when it was appointed coincident with the proclamation of the Power Corporation Act on 4 March 1974 was that four of the seven Task Force Hydro members – Taylor, Muncaster, Gordon, and Frame – were among the thirteen new directors. Indeed, Taylor was made vice-chairman and chairman designate, and Gordon was made president. In a context where two of the remaining three task force members, Reynolds and Dillon, were deputy provincial secretary for resources development and deputy minister of energy, and McKeough was minister of energy, the government was clearly seeking to implement the task force's remaining recommendations through appointments. Dillon remarks that appointments constituted the 'implementation phase,' stressing that it was important to appoint people who believed in the task force's recommendations. Similarly, Gordon recalls that with their head start, the task force members were 'able to move right in and perform their duties.' Davis does not recall any plan to use appointments to accomplish what had not been accomplished by statute. Rather, he believes that the task force members could have been appointed in their own right but were appointed to see through the recommendations of the task force because of their knowledge of the intricacies of Ontario Hydro.

The remainder of the board appointees reflected a mixture of broad interest and geographic representation. The new directors were Arthur Evans MPP, formerly vice-chairman of the HEPC, Bradford; Douglas Hugill, president of the OMEA 1969–70, Sault Ste Marie (who with Frame made two OMEA representatives); Allen Lambert, chairman and CEO, Toronto Dominion Bank, Toronto; Conrad Lavigne, president, Mid Canada Television System, Timmins; Philip Lind, chairman, Sierra Club of Ontario, and secretary, Rogers Cable, Toronto; Jean Pigott, president, Morrison-Lamothe Foods, Ottawa; Robert Uffen, dean of applied science, Queen's University, Kingston; and the thirteenth, appointed later in 1974, was W. Dodge, formerly secretary-treasurer, Canadian Labour Congress, Ottawa. Thus, whereas the Ferguson formula had limited HEPC appointments to government and OMEA representatives, Ontario Hydro's representation was much broader. It now included various sectors of the business community, a business woman, an environmentalist, and a labour leader.

In accounting for the long delay between the passage of the Power Corporation Act in June 1973 and the appointments in March 1974, McKeough recalls that it took a long time to put the new board together. He nonetheless concedes that time was needed to 'let the dust settle' on the select committee report on Hydro Place, released 2 October. The lag was caused in part because Gathercole, who believes his relationship with Davis 'cooled' after the Hydro Place affair and that the affair precluded him from continuing as chairman, refused to retire before he turned sixty-five in December 1974. Believing the select committee cleared Gathercole of wrongdoing, Davis does not recall the affair causing him this discomfort or compromising his tenure, but concedes that a consideration of personalities might have been a factor in delaying the appointments. McKeough, however, reveals that, as minister of energy, he had insisted on Gathercole's departure before reversing himself and letting Gathercole stay on for the remainder of 1974 because the chairman had said the 'right things' at the select committee hearings.

7

End-of-the-Century Public Ownership: An Epilogue for Ontario Hydro, 1974–95

While another book could be written on the history of Ontario Hydro since 1973, the themes of the foregoing story can be succinctly pursued so as to encompass its development since. The most significant feature of Hydro's reconstituted relationship with the Ontario government was that the informal, political nature of past relations, largely built on the Ferguson formula for appointments, was replaced by a more formal, statutory relationship. Although this scheme was intended to lessen the problems Hydro posed for the government, it had the opposite effect. Hydro's second-generation crown corporation plans for a massive expansion of capacity were immersed in public scrutiny. There was no returning to the informal relationship, but the grafting of its strength – the strategic judicious use of appointments – onto the new formal relationship proved to be an improvement. In no small measure, this helped to reduce controversy over Hydro and, in response to the recession in the early 1980s, enabled Hydro to begin the transition to a third-generation crown corporation. While this process was interrupted before the decade had ended by a plan for new expansion, the more protracted economic downturn of the early 1990s has hastened the transition from social purpose to business logic.

Formalized Government-Corporation Relations

The politics of the post-HEPC era began with Darcy McKeough, the minister of energy, rather than Bert Lawrence, the provincial secretary for resources development, having taken charge of the selection of Ontario Hydro's board of directors. While this was not a surprise, the event revealed that Lawrence had not only been shut out of Hydro's affairs, but was also sidelined by the new cabinet structure generally. After he resigned on 26

February 1974, the week before the directors were appointed, Allan Grossman, clearly then a junior minister to McKeough, was appointed in his stead. Thereafter, the policy minister never enjoyed the senior coordinating status that the Committee on Government Productivity had intended.

While the government had sanctioned Ontario Hydro's ambitious expansion plans in June 1973, they had become sufficiently controversial for the government to appoint a royal commission on electric power planning, with Dr Arthur Porter as chair, on 13 March 1975. The controversy nonetheless moved ahead of the government when Hydro proposed on 24 April a 29.7 per cent rate increase for 1976, more than double what it had accepted after the first Ontario Energy Board review for 1975 rates. Before the increase could be steered through the board rate hearing, it faced political scrutiny. On the eve of an election campaign, McKeough, who was reappointed treasurer on 18 June, used a mini budget in July to instruct Hydro to reduce its capital expenditures by $1 billion and reconsider the amount of its rate increase. This action confirmed that his replacement as energy minister, Dennis Timbrell, until then a minister without portfolio, would be junior in the cabinet. In response to the new financial constraints, Hydro chairman Robert Taylor, with the election looming, revised the increase downward to 25 per cent on 31 July,[1] but all was for naught as the government was reduced to a minority in September.

The release of the energy board's rate recommendation on 10 October 1975 revealed that the Task Force Hydro reforms had not served to distance the government from Ontario Hydro's affairs. The problem arose because the board recommended that Hydro levy a 27 per cent increase, two points more than proposed, to ensure financial stability. The recommendation, which was followed on 14 October by the unveiling of the federal anti-inflation program of wage and price controls, left Premier Davis needing to address the opposition and public outcry. With the Ontario Commission on the Legislature having recently recommended more responsibilities for members, he established the Select Committee on Hydro Affairs. Since the government had only a minority, former NDP leader Donald C. MacDonald was appointed chairman. In its interim report of 17 December, the committee recommended a 22 per cent rate increase for 1976, which Ontario Hydro, still with final approval over rates, accepted on 19 December.[2] This legislative scrutiny later spiralled into a general assault on the underlying assumptions of Hydro's expansion plans.

On the heels of the rate controversy, McKeough did what no treasurer had ever done publicly. On 16 January 1976 he threatened to withhold the government's guarantee of future Hydro debt if the corporation did not reduce the growth of its annual expansion from 7 to 5 per cent. Added to this broadside, the select committee, believing that rates were not a separate issue, recommended in its final report that June that Hydro adopt conservation and demand management policies.[3] Given that the Porter Commission had now begun public hearings, the deluge did not stop, but little political support was forthcoming from the government because of its minority status. Moreover, the energy ministry changed hands again on 3 February 1977 when Timbrell, elevated to minister of health, was replaced by James Taylor, the former minister of community and social services.

Although the government failed again to win a majority in the June 1977 election, in part because of Ontario Hydro's problems, it had not yet given up on Hydro's expansion plans. They remained an important countercyclical economic stimulus for the government, consistent with Hydro's second-generation crown corporation status. Hydro was in fact pump-priming the economy by $1.5 billion per year through its capital expenditures. Unwilling to bring this stimulus to an end, the cabinet gave its approval on 18 July for the construction of the Darlington nuclear station, then estimated to cost $4.5 billion, and exempted the project from the new Environmental Assessment Act. The government could not, however, preclude the re-establishment of the select committee. On 24 November it was given a new, broader mandate, including an investigation of Hydro's commitment to nuclear power, and thereby appeared to duplicate the task of the Porter Commission.[4] The capacity to navigate Hydro issues through the minefield of these various points of scrutiny was hampered again when Reuben Baetz was appointed from the backbench to replace Taylor in the energy portfolio on 21 January 1978.

The exposure given to the expansion program by the energy board, the Porter Commission, and the select committee revealed that Ontario Hydro had built huge excess capacity into its system with more yet to come. As the government had encouraged expansion but had not come to Hydro's defence over the issue, the corporation formally asked for policy direction on 10 April 1978. Baetz replied that Hydro should return to water power, a politically convenient but abrupt change in policy. He was not around long enough, however, to see the change through, having been replaced by the more experienced James Auld, the former chairman of the management board, on 18 August. Complicating matters further, the Porter Commission released an interim report in September which concluded that electricity demand growth would be only 4 per cent rather than the 7 per cent Hydro had anticipated. Given this circumstance, Porter recommended, like Baetz, that Hydro pursue more flexibility in power production and not rely exclusively on large-scale nuclear plants for new capacity.[5] This view, expected to be amplified in the final report, meant Hydro policy was now a major political problem for the government.

Since the government's capacity to shelter Ontario Hydro from scrutiny had been lost in the new formal, statutory relationship, its only option was to remake the relationship through a more strategic and judicious use of appointments. The transformation began when Hugh Macaulay, Robert's brother, was appointed to the Hydro board of directors as chairman designate on 1 January 1979. He was subsequently elected vice-chairman of Ontario Hydro and chairman of the social responsibility committee at his first board meeting, and he assumed the new tasks on a full-time basis much like his counterparts under the old HEPC structure. Hugh was himself an important political confidant of Davis, having been chairman of the Ontario Progressive Conservative party from 1971 to 1976. He became Hydro chairman on 1 July when Taylor stepped down six months before his five-year term expired.

On the government side of the relationship, Robert Welch, the deputy premier and provincial secretary for justice, was appointed minister of energy, replacing Auld on 30 August 1979. Although he remained deputy premier, his appointment was not a demotion. The premier had chosen to assign cabinet talent and senior status to the portfolio, acknowledging the government's past failure to manage Hydro's affairs through appointments. Welch was the sixth energy minister appointed in the six years since the creation of the portfolio in 1973. Through the informal nexus of Macaulay and Welch, the formal, statutory relationship came to be managed through a common interest reminiscent of the HEPC before 1974, where Hydro's need for business autonomy was mediated with the government's desire for policy sensitivity. The focus of Macaulay's and Welch's attention was the Porter Commission and the select committee. While the latter presented many political challenges, the former, which was due to deliver its final report in February 1980, was the more immediate problem.

The Porter Commission's recommendations were an indictment of the expansion program and, counter to Davis's new emphasis on appointments, the incomplete nature of the formal, statutory relationship. The commission recommended that the rigidity of supply planning, with its fixation on large-scale nuclear plants, be abandoned for the flexibility of demand management and smaller-scale additions to generation capacity. Moreover, it doubted whether the certainty of the past, which had justified supply planning from Ontario Hydro's inception, would ever return. On the formal, statutory relationship, it recommended that the Ontario Energy Board be transformed into an Ontario Energy Commission to integrate all energy policy-making in an independent body. The Ministry of Energy accepted eighty-two of the commission's eighty-eight recommendations, rejecting the four that would have further institutionalized the formal, statutory relationship.[6] For the government, the political wisdom of a strengthened informal relationship had shone through and paid dividends, so much so that there was little political fallout from the commission's exposure of Hydro's problems.

The Select Committee on Hydro Affairs, the other challenge for Macaulay and Welch, was highly sensationalist in its orientation, particularly after the nuclear accident at Three Mile Island, Pennsylvania, in March 1979. While there was little the minister and chairman could do but ensure good public relations and rebuild the government's and Hydro's image on energy matters, this strategy turned out to be a success. Indeed, by the time of the March 1981 election, the controversy over Hydro had died down sufficiently for the government to include accelerated construction at Darlington and an increased reliance on electrical energy in the province in its campaign platform.[7] As the Conservatives were returned with a majority after six years of minority government, the era ended when select committee became a celebrated victim of the outcome.[8]

A Third-Generation Crown Corporation?

Ontario Hydro's nationalistic nuclear industrial strategy collapsed in 1982 under the weight of recession and conservation, bringing an end, for the time being, to its second-generation crown corporation mission. Given new international constraints on borrowing and high interest rates, all debtor governments were forced to manage their public sectors better. Hydro did not escape this circumstance. Although the government had amended the Power Corporation Act on 26 June 1981 to include energy conservation as one of Hydro's purposes, the corporation still sought to legitimatize the expansion program by exporting excess capacity. Even this strategy failed when the American recipient, General Public Utilities, ended the contract on 11 June. Hydro extricated itself from its predicament by quietly adopting the fortuitously delivered demand management recommendations of the Porter Commission, a decision manifest in the slowdown of construction at Darlington and other projects and the cancellation of still others outright in 1982.[9] In doing so, Hydro took its first steps towards becoming a third-generation crown corporation, one where business logic replaces social purpose as the operative principle. Hydro could make the transition only imperfectly, however, because of the high visibility of its monopoly status and statutory existence.

The transition was aided by the Supreme Court of Canada's decision in the case of *Nepean* v. *Ontario Hydro* in 1982, which further undermined the statutory basis for the contention that Hydro was the trustee of a municipal cooperative. In 1974 Nepean had begun to withhold its net contribution to Hydro's 'return on equity/cost of return' accounting system for the sinking fund payments. It also launched a suit to recover past payments on the grounds the charges were not a cost of power as prescribed in the Power Corporation Act, whereupon Ontario Hydro also made a counter-claim for the payments since 1974. In the majority decision, Justice Willard Estey did not consider a determination of whether the municipal hydros had 'acquired an equity in law or fact in the assets of Ontario Hydro' essential to the adjudication of the case. He closed, however, with the aside that 'Each party is a public body "owned" by the people of the Province in the one case and by the township [of Nepean] in the other.' Justice Brian Dickson, in the minority decision, addressed the cooperative contention directly. He accepted the lower court's view that the municipal hydros 'did not acquire an "equity" in ... Ontario Hydro' and that even if they did the equity would be 'vague and uncertain' and would not convey any rights unless Hydro ceased operation. Thus, while the *Nepean* case primarily hinged on monies paid under mistake of law, with the court deciding in Nepean's favour but with neither parties' funds recoverable, the case nonetheless had implications for who owned Ontario Hydro. The ambivalence remained, but the basis for the cooperative contention was narrowed.[10]

With Hydro issues sufficiently depoliticized in the aftermath of the 1981 election, political talent of the calibre of Macaulay and Welch was no longer needed to manage the formal, statutory government-corporation relationship. Macaulay resigned on 31 March 1983, leaving President Milan Nastich, who had succeeded Douglas Gordon in 1980, to serve also as acting chairman. Welch relinquished the energy portfolio to Philip Andrewes, a new minister, on 6 July, but remained deputy premier. Thereafter, Davis struggled with the matter of appointing a new chairman, a decision intertwined with his own retirement plans. Rumours circulated that he was saving the post for his long-time deputy minister, Edward Stewart, as a reward for service. When Davis's departure was delayed into 1985, Tom Campbell, a respected senior civil servant, was appointed chairman on 20 August 1984. The new appointments reflected the successful arrangement of the 1960s where the government, with a majority, had George Gathercole, a former senior civil servant, in the chairman's post and a junior minister in the energy portfolio. In such a circumstance, the minister provided a public face for Hydro policy, enabling the premier to maintain direct but informal relations with the chairman.

The return to the old ways was interrupted by the May 1985 election. Having won the most seats, Frank Miller, Davis's successor as Conservative leader, formed a minority government with Mike Harris as minister of energy, but it was soon defeated. In its place, the Liberals, who had signed a two-year legislative accord with the NDP, were called upon to form a government without another election. Although out of office since 1943, they had triumphed in the long battle with the NDP to succeed the forty-two year Conservative dynasty. The change of government left Campbell, who had taken a leadership role in the Conservative campaign and previously been deputy treasurer under Miller, in an awkward situation.[11] With no clear precedent for his removal, given only four changes of government since Ontario Hydro's inception, the new premier, David Peterson, did not force the issue. Although he had been highly critical of Hydro's operation and direction, his only immediate action was to appoint Vince Kerrio, his party's long-time energy critic, to the energy portfolio. By not appointing a senior minister, Peterson appeared to reserve for himself the informal channels of influence in Hydro's affairs that the Conservatives had so long cultivated but only recently revived.

Campbell could remain, and the government be spared the potential fallout from his removal, because Peterson decided that a shakeup of Ontario Hydro could be better performed by other means. When Nastich resigned as president on 31 December 1985, Peterson let it be known that he wanted Robert Franklin, the executive vice-president of CN Rail, to be Nastich's replacement. The board of directors, whose statutory prerogative it was to make the appointment, complied with the wish in January 1986. This political intervention was significant. It was the first time since Hepburn had become premier in 1934 that the government selected the chief operating officer. In this regard, the Peterson government overstepped a division of labour in the Ferguson formula that distanced the government politically from Hydro's operation. The intervention was also the first time ever that the chief operating officer had been chosen from outside the corporation. Lastly, it was the first time the post was not held by a professional engineer. The appointment appeared to consign Campbell to the routine affairs of the board of directors, despite being full-time, and designate Franklin to be the conduit of the informal relationship to the premier and the formal relationship to the government. Ironically, it was in this environment that Hydro's expansion program was renewed. The economy had come out of recession, so in August the government permitted the full completion of Darlington, now an $11 billion project. And Campbell, although rebuked by Peterson, publicly speculated on the need for another large nuclear station in November.[12]

Peterson's control over Hydro issues was greatly enhanced when he won a majority government in September 1987. He again maintained for himself an informal relationship with Franklin by appointing Robert Wong, a rookie MPP, as energy minister on 29 September in place of Kerrio, now minister of natural resources. He then announced through the 4 November speech from the throne his government's plan to refashion its relationship with Hydro. The speech promised 'greater public accountability and responsiveness' by Hydro through amendments to the Power Corporation Act and the appointment of a select committee on energy. Consistent with the division of labour already established between Peterson and Franklin, there were indications that the president would be made the CEO, with the chairmanship reduced to a non-executive part-time position. Faced with this diminution of responsibility, Campbell resigned on 18 January 1988, three and a half years into his five-year term. His decision was no doubt influenced by the fact that Peterson had appointed Campbell's wife, Mary Mogford, as deputy treasurer. If he stayed, Campbell would have been in an untenable situation, given Hydro's close financial relations with Treasury.[13]

While remaining president, Franklin was appointed chairman. Thus, the division of labour between the chief executive and chief operating officer that was as old as the HEPC and Ontario Hydro was removed. This division effectively insulated the government from direct political responsibility for Hydro's day-to-day operation while allowing the government political influence in its affairs. The chairman, as the government's appointee, had always mediated the government's and the corporation's competing needs for policy sensitivity and business autonomy and could do so by not also being charged with the day-to-day management of the corporation. This dual role was compromised in Franklin's case. Although Thomas Hogg and Nastich had held both posts simultaneously, their cases differed in that, as career Hydro employees, they did not come to the utility as political appointees. However, barring a major political gaff, the absence of the division of labour could be managed as long as the Liberals remained in power.

The government moved to implement its vision of the government-corporation relationship through two pieces of legislation in 1989. In response to a provision of Robert Nixon's budget that year, the Power Corporation Act was amended in June to charge Ontario Hydro annually for the financial benefit of the government's debt guarantee. The fee, to be equal to one-half of 1 per cent of the debt, was projected to be $138 million in 1990–1.[14] Although this brought to an end the government's historic commitment to use its financial strength to aid Hydro as a facilitative crown corporation, the intent was to put Hydro on a level playing field with other energy producers. This change was consistent with the business logic of a third-generation crown corporation. Following in the same vein, the act also received two notable amendments in October. One broadened the definition of conservation established in 1981to include the reduction of electricity use. The other, in the first-ever privatization thrust, added the promotion of non-utility generating companies, or NUGs, to the purposes of Hydro.[15] Before the latter amendments passed, Lyn McLeod, until then the minister of colleges and universities, replaced Wong as minister of energy on 2 August.

Other amendments to the act in October further formalized the statutory relationship. While the changes were consistent with the Liberals' long-held view that Hydro had been out of control, they were inconsistent with the flexibility required of a third-generation crown corporation. Every three years the minister of energy and Hydro were to produce a 'memorandum of agreement' on matters of policy and accountability to the minister. In addition, the minister was given the power to request that Hydro submit plans for scrutiny; to issue policy statements to which Hydro was required to report on implementation; and to appoint inquiries under the Public Inquiries Act. The last two would need prior cabinet approval. And finalizing what Peterson had already established through appointments, the president was statutorily declared the chief executive officer. To make this possible, the newly titled chairperson's designation as chief officer, along with the requirement that the chairperson work full-time, were deleted. The Power Corporation Act was also amended to increase the size of the board to seventeen from thirteen directors and remove the limit of three terms of three years for ordinary directors.[16] This all left Hydro more closely tied to the government and the government with more direct responsibility for Hydro's affairs through Franklin, with the result that the corporation had less rather than more flexibility to operate outside the government's economic objectives.

The Liberals promoted their refashioned relationship ostensibly to discipline Hydro to the conservation and environmental agenda they had taken since forming the government in 1985. However, in December 1989, only two months after restructuring the relationship, Franklin forwarded to the minister of energy an ambitious expansion plan for review by the Environmental Assessment Board. Under the rubric of a twenty-five-year demand/supply proposal, Hydro planned to build a new large-scale nuclear generating station at Blind River on the north channel of Georgian Bay. This return to supply planning was softened by being joined with demand management programs. Although Franklin stated that there would be power shortages without the additional power, the wisdom of this industrial strategy, reminiscent of a second-generation crown corporation, was in doubt as early as the spring of 1990. By then the economy was heading for recession. The outlook was so bleak that the Peterson government, only three years into its mandate, took advantage of its high standing in the polls to call a September election before it could be held responsible for the province's economic woes.[17] This strategy backfired. The NDP, which had never held power in Ontario, won a majority government, inheriting the Liberals' new Hydro relationship and the expansion plans in a vastly changed economic environment.

Restructuring and Privatizing Ontario Hydro

On taking power, the new government gave two signals that Ontario Hydro's expansion program was in jeopardy. The premier, Bob Rae, appointed Jenny Carter, an environmentalist, as minister of energy on 1 October 1990, and the speech from the throne on 20 November announced that the government would place a moratorium on new nuclear development and intensify energy conservation and efficiency efforts. In such a circumstance, Franklin's tenure as chairman and president, given his support for expansion and his close connection with Peterson, could not survive. When Franklin stepped down at the end of May 1991, Rae appointed Marc Eliesen as chairperson on 24 April, subject to a review hearing before a legislative committee. Eliesen, who had been chair of Manitoba Hydro under the NDP and federal NDP research director, was the deputy minister of energy who had been brought into the government by Peterson. Knowing of Rae's intention to make him the CEO as well as chairperson, contrary to the Liberals' 1989 statutory changes, the board of directors took two actions hostile to the government: it informed Rae that Eliesen was not qualified to be the CEO, and elevated Alan Holt to president and CEO from vice-president corporate planning on 14 May. As a result, Eliesen, whose committee hearing was not held until 5 June, was not permitted a role in the selection of such a close working associate.[18]

The government responded to the board's actions by introducing amendments to the Power Commission Act on 5 June. Among other objectives, these amendments reversed the statutory changes made by the Liberals. In the process, the executive structure was left equally inflexible compared with what the Liberals had established. The bill, after changing all chairperson references, declared that the chair, rather than the president, was to be the CEO and was to be full-time, an arrangement that was to be retroactive to the introduction of the bill. The legislation, which was passed on 25 June 1992, also increased the size of the board of directors to twenty-two from seventeen. The additional five were to consist of the deputy minister of energy as an ex officio non-voting director and four regular directors, presumably to permit the NDP to further its conservation agenda. While the former addition was a scheme first recommended by Task Force Hydro in 1972, it was now implemented, ironically, to ensure the government was informed of Hydro's plans rather than the reverse. In addition, the act would now permit the minister to issue policy 'directives,' rather than 'statements,' but with the new provision that the minister must consult with the board first and that the directives were to be made public in the *Ontario Gazette*.[19] Despite the greater visibility of government in this scheme, the amendments put Hydro back on the commercial path of a third-generation crown corporation because government policy could no longer be dictated to the corporation and hidden from public view.

Having won the battle with the board of directors, Eliesen immediately consolidated power over policy in his office. His executive reporting structure placed six operational vice-presidents under Holt and had Holt and five policy vice-presidents, such as finance and corporate planning, report directly to him. Rae helped as well by appointing Will Ferguson, a brusque backbencher, to the energy portfolio on 31 July in place of the meek Carter. On 15 January 1992, only six months later, Eliesen applied to the Environmental Assessment Board to defer the large-scale generation component of Franklin's demand/supply proposal. Although the reason given was weak demand for power in a recession economy, this action precipitated the plan's later demise. In lieu of building this new capacity, Eliesen planned to refurbish Hydro's existing generating stations, nuclear included, to be financed by a rate increase of 44 per cent over three years.[20] Although he felt this scheme would not have a deleterious effect on the Ontario economy, it led some private industries and municipal hydro commissions to consider generating their own power with natural gas, which now had a price advantage.

Despite the turnaround of the expansion plan, the leadership that brought it forward soon came to an end. Ferguson, although later cleared of the charges, resigned from the energy portfolio on 13 February as a result of a police probe of a reform school for young women where he had worked as a young man. He was replaced by Brian Charlton, the minister of financial institutions, who had originally been left out of the cabinet for the presumptuousness of having made a policy pronouncement as the rumoured minister of energy. Eliesen, only fifty-one, then announced on 31 July that he would resign at the end of October to become CEO of British Columbia Hydro. His departure left the ominous message that Ontario Hydro's problems were too deep for him to sort out. Before he left, his nemesis as president, Holt, retired in late September after reportedly receiving a 'golden parachute.'[21]

With Hydro's leadership slate cleared, Rae appointed Maurice Strong as chair and CEO of Ontario Hydro and the board of directors later named Alan Kupcis, then vice-president of procurement and power system planning, as president and chief operating officer. Strong, an internationally respected business leader who had just chaired the United Nations' earth summit in Rio de Janerio and was a family friend of Rae, had the right credentials to mediate the corporation's and the government's respective needs for business autonomy and policy sensitivity in the current climate. Despite nearing retirement at sixty-three, he acted boldly. The signal of the whirlwind of change ahead came when he declared Hydro 'a corporation in crisis' before the legislative committee reviewing his appointment on 9 December 1992. Hydro's revenues had been severely undercut by the recession, so his first priority was to reduce operating costs. He began on 11 December by offering a voluntary severance package designed to cut Hydro's 29,000 staff by 2,000. The cancellations of numerous capital projects and purchase agreements followed. These included power contracts with Manitoba Hydro, at a penalty of $150 million, and small non-utility generating companies, both justified by a surplus of generating capacity. And bringing the 1989 expansion plan to an end, Hydro withdrew its demand/supply proposal from the Environmental Assessment Board on 25 January 1993.[22] To complement Strong's orientation, the ministries of the environment and energy were informally amalgamated in February under the new leadership of Bud Wildman as minister.

Strong's plan for a radical transformation of Ontario Hydro was outlined that February in a position paper entitled *Hydro 21*, but it was not a complete departure from the past. By maintaining that Hydro had 'a crucial role to play to help Ontario compete economically,' although now in 'an environmentally sustainable way,' the paper still considered sacrosanct Hydro's historic facilitative crown corporation role in the provincial economy. Yet there was a difference under Strong in how this role was mobilized. In place of the social purpose of a second-generation and nationalistic crown corporation that had been the guide in the 1970s, Hydro would now be given the flexibility to complete its transition to the business logic of the third-generation. To this end, the document posed a range of structural reform options, from Hydro's remaining 'vertically-integrated and centrally-coordinated' to its being replaced by a loosely connected system of 'small, specialized, competing organizations.' Moreover, ownership options along the range could vary among public, private, and mixed.[23] After eighty-seven years of being a public institution, Ontario Hydro had begun to consider the merits of privatization, if only in part.

Along the road to his restructured Ontario Hydro, Strong pursued the cooperation of the Municipal Electric Association, successor to the OMEA, for his restructuring initiatives. In his speech to the association's convention on 1 March 1993, he offered to 'forge a new era of partnership,' hoping to stem the tide of disenchantment over expansion and rates. Despite the increase in the size of Hydro's board to twenty-two members from the thirteen in 1974, municipal hydro representation remained fixed at two directors. Thus, one 'fundamental' change Strong proposed was additional appointments. Other points were given broader dissemination in his 9 March restructuring announcement entitled *Hydro in Transition*. Foremost, this document spoke to his desire to reduce staffing levels by 4,500 over and above the 1,500 employees previously enticed to leave. Beyond a reduction of operating, maintenance, and administrative costs by 25 per cent, Strong proposed to terminate capital expenditures on Eliesen's plan to refurbish existing generating stations, with the admission that the Bruce A nuclear station would now operate while 'safety permits.' In order to improve the revenue picture, Strong made a commitment not to increase rates for 1995 and to freeze them at the level of inflation for the rest of the decade. By making Hydro more 'cost effective, accountable and market-oriented,' the scheme was designed to dissuade existing municipal and industrial customers from generating their own power and adding to the revenue loss.[24]

The restructuring plan and a corporate reorganization were approved by Ontario Hydro's board of directors on 13 April 1993, and released publicly in a pamphlet titled *The New Ontario Hydro*. The executive reporting structure now divided Hydro into three autonomous business groups, each headed by a managing director. These managers, who did not sit as directors by statutory provision, were to report through the chief operating officer to Strong as chief executive officer, who was also supported by corporate-wide finance, human resources, and public affairs functions. In practice it was an informal circle. The first of the three groups was the Electricity Group, the core business, assigned to Kupcis; the second was an Energy Services and Environment Group, assigned to John Fox, an executive hired away from Pacific Gas and Electric of San Francisco, to provide customers with management services; the third was an Enterprises Group, assumed by Strong. This last group was charged with turning a profit on Hydro's technologies and expertise, and it exemplified the new commercial emphasis. While the board did not also forward a scheme for privatization, it established 'equity of access and price' as strategic principles for the electricity business, and thereby established a broad but well-defined parameter for debate on privatization.[25]

Despite the restructuring and reorganization, the board's mission for the whole corporation was clearly still facilitative, although it now included Strong's commitment to sustainable development. Maintaining a long-held belief, the board adopted the view that 'Hydro's own competitiveness will contribute to the competitiveness of the provincial economy.' However, the board's interpretation of the mission, including sustainable development, led to two political controversies in the summer of 1994. First, after facing a storm of opposition, Hydro dropped plans to offset its carbon dioxide emissions by purchasing portions of a Costa Rican rain forest. Secondly, Ontario Hydro International Incorporated, Hydro's new subsidiary, purchased a stake in Edelsur, an electrical utility in Lima, Peru. In response to the latter, the cabinet referred Hydro International's activities to the Ontario Energy Board for review. In its April 1995 report, the board put a damper on such purchases, recommending that Hydro confirm that the primary objective of Hydro International is to provide benefits to Ontario ratepayers and that Hydro place a 'definite cap' on its subsidiary's use of ratepayer-derived funds.[26]

Continued restructuring and some form of privatization were obviously on the horizon, so the Municipal Electric Association (MEA) began to examine the 'institutional options' for reform of the electric system in Ontario. In its 27 May 1993 Phase I report on alternatives, it determined that while transmission and distribution were natural monopolies that should remain in the public sector, there was no technological need for a monopoly on generation. This made the main issue for reform an unbundling of the vertical integration of generation, transmission, and distribution, together with the related issues of privatization of generation and third-party access to transmission, known in the industry as wholesale wheeling. With generation the focus of reform, the options the MEA enumerated, in addition to public monopoly, were managed competition and open competition. Adding a refrain not heard for many years, the association stated that Ontario Hydro's ownership would need to be resolved before any privatization could proceed. Moreover, it pointed out that the municipal hydro commissions had 'never relinquished their claim to [Hydro's] ownership.'[27]

Although Wildman declared in July 1993 that he was opposed to privatization, the MEA's Phase II report on alternatives continued to pursue this end when completed on 19 August. In a scenario where Hydro's generation assets were unbundled from transmission, the association's task force felt the transmission network should be taken over by its present participating municipal hydros and new rural power cooperatives operating in place of direct distribution by Hydro. This new transmission cooperative would purchase power in a framework of managed competition during a transitional period of seven years so as not to strand Hydro's existing generating assets through open competition. Although Wildman was prepared to consider joint public-private ventures, he informed Hydro the next day that the government saw no public policy benefit in privatization. Indeed, except for Energy Probe, a neo-liberal environmental group, he saw no political support for privatization. The Association of Major Power Consumers, successor to the Niagara Basic Power Users, feared it would lead to higher rates.[28]

At the February 1994 MEA convention, Strong sought to divert the association's attention from a privatization of generation. Facing the municipal challenge directly, he argued that 'privatization is not the issue – the issue is the financial integrity of the corporation.' The answer, for him, lay in making Hydro more businesslike, with one means being to build on the Hydro-MEA partnership of old. Acknowledging the historical basis for the municipal ownership claim, he exposed its mythical foundation, stating that 'the current reality is that there is no legal basis for this position, and at present you exercise none of the prerogatives of ownership.' Falling back on the safety of the institutionalized ambivalence so long promulgated to diffuse this contentious issue, he submitted that the claim 'should not be and need not be dealt with as a legal issue, but rather through the political process.' He included here, as he had in 1993, increased municipal hydro representation on Ontario Hydro's board, although a third representative would not be appointed until 1995. Arguing that the MEA's preference for unbundled generation served its own institutional interests, he stated that their partnership entailed an 'obligation to ensure that the equity already in the system [was] employed to the greatest advantage ... before seeking equity infusions from outside the system' through privatization.[29]

Unbeknownst to the MEA, its ambition for a municipal takeover of Ontario Hydro's transmission system was undercut by a redeployment of equity orchestrated by Strong. Hydro's 1993 annual report, released 14 March 1994, revealed a $3.6 billion loss, the largest in Canadian corporate history. The figure, for the most part, was accounted for by one-time restructuring charges, including the separation costs for 6,700 regular and

4,000 contract employees. The write-off, according to Eleanor Clitheroe, Hydro's senior vice-president and chief financial officer, was a 'normal commercial practice.' By being taken out of equity, however, it severely diminished the accumulated debt retirement appropriations of the municipal hydro commissions. This fact was not immediately apparent. The board of directors had decided on 13 December 1993 to consolidate the annual report's previous presentation of separate equities into the one heading of 'retained earnings,' which brought to an end the practice begun in 1916 of itemizing municipal equity in Ontario Hydro. In the 1992 report the debt retirement appropriations had represented $5.2 billion of the $6.9 billion in equity, but the 1993 report listed retained earnings as just $3.3 billion, and thus less than the previous total of municipal hydro equity. In doing so, Hydro had reduced its other equities – contingency reserves of $1.6 billion and government contributions regarding rural electrification of $127 million – to zero. This meant the municipal appropriations absorbed the $1.9 billion remainder of the $3.6 billion write-off and became the only equities in the retained earnings. As a percentage of Hydro's assets, they were lessened to 7.4 from 11 per cent at the close of 1992.[30]

The MEA's interim Phase III report on implementation was completed on 31 March 1994 but was not drafted to take into account how the municipal hydro equity in Ontario Hydro had been redeployed. Its task force recommended that municipal and new rural hydro commissions should take ownership of Hydro's transmission system by acquiring the book value of the debt assigned to these assets, with shares divided on a pro rata equity basis. Moreover, these municipal and rural 'owners' would appoint its board of directors, to which their 'commissioners and managers' would form the strongest representation. This control of the company was what OMEA/MEA leaders had always wanted to supplement the claimed municipal cooperative ownership of Hydro, unsatisfied as they had been with influence garnered in the Ferguson formula for appointments. The workability of the plan, however, was based on the premise that the combined equity of the municipal hydro commissions and their equity in Hydro was enough to cover the purchase of the transmission network. With the viability of the plan affected by the $3.6 billion write-off, the MEA's final Phase III report, released in September 1994, recommended only that the transmission company be 'publicly owned.'[31]

Although Strong had chided the MEA for pursuing its own self-interest, Ontario Hydro was pursuing the same. While the 27 June 'challenges and choices' report of the utility's financial restructuring group acknowledged that the status quo was no longer a viable option, it also recognized that opening access to Hydro's transmission lines – an action which would include American producers as well as Ontario-based NUGs given the Free Trade Agreement – would strand generation assets. The solution to this predicament lay in instituting transitional arrangements, including compensation from generating companies granted access to the transmission system. To strengthen Hydro's position in such a managed competition scenario for generation, the report held that a share offering would be a financially prudent route to improving Hydro's debt/equity ratio, made worse by the write-off. Speculation ran that this might generate $5 billion. Although the report conceded that such a privatization would require enhanced regulation, it stated that Hydro would need increased freedom from the constraints of the Power Corporation Act, a key requirement of a third-generation crown corporation. The act's historical interpretation in support of uniform pricing, for example, ruled out the flexible pricing now permitted in some jurisdictions for retaining large customers. Before such a privatization should proceed, however, the report recommended that Hydro's ownership be clarified, noting that the government's controls were, 'in business terms, clear indications of ownership.'[32] The statute nevertheless still required municipal debt retirement payments.

An outline of the partially privatized Ontario Hydro appeared in an internal document entitled 'The Commercialized Hydro' in July 1994. Although a think piece as opposed to an action plan, it favoured a gradual movement of Hydro out of generation. In the future Hydro would market its engineering and management expertise to independent producers to help them bring power on stream rather than build its own stations. In this scheme of managed competition, Hydro would remain in control of the transmission network, recognizing that this would be essential to balancing the competing pressures for third-party access to the system and ensuring its generating assets were not stranded.[33] While the government had stated publicly that it did not favour privatization, Strong's plan for a gradual breakdown of Ontario Hydro's monopoly in exchange for managed competition was not inconsistent with Wildman's policy pronouncements. As the guarantor of Hydro's debt, the government had a vested interest in Strong's success.

Now that Ontario Hydro had made great strides reorienting itself to the commercial objectives of a third-generation crown corporation, Strong was able to announce the first rate decrease in thirty years on 17 October. Rates for 1995 would be cut by 0.7 per cent for large industry and frozen for all other customers. He added that Hydro was now in the recovery room, with the sense of major crisis overcome. With the worst of his task behind him,

Strong had reduced his leadership commitment to that of a 'dollar-a-year' part-time chair in September. Although he remained CEO, Hydro's three main operating divisions now reported to Kupcis. Publicly, Strong stated that he wanted the freedom to pursue other interests, but privately it was suggested that he wanted no impediment if called upon to become secretary-general of the United Nations. To legitimate the change, the Power Corporation Act was amended on 8 December 1994, removing the requirements that the chair work full-time and also be the CEO. The cabinet was instead given the flexibility to appoint either the chair or the president as CEO, but had to consult the board of directors first. Kupcis was appointed CEO on 17 January 1995.[34]

All that remained unresolved from the Strong era was privatization. In Ontario Hydro's 1994 annual report, Strong and Kupcis mused on the deregulation of the energy industry in North America, arguing that 'Hydro can survive in the long-term only if it continues to adapt aggressively, in a businesslike fashion, to an increasingly open and competitive environment.' One notable change instituted in Hydro's transition from a basic commodity supplier to a customer-driven business has been internal competition among generating units to increase efficiency. While such continued restructuring was considered admirable by the Rae government, privatization gained new impetus with the outcome of the June 1995 election. The Progressive Conservatives, who won a commanding majority under leader Mike Harris, promised moves toward privatization in their platform of a 'Common Sense Revolution.'[35] Strong's restructuring initiatives have facilitated this end by severely diminishing the equity basis for the myth of municipal cooperative ownership. This now faint notion, however, still poses a political complication for privatization because of the vigour with which it is held by municipal hydro leaders. The institutionalized ambivalence of Ontario Hydro's ownership in statutes and appointments, not yet dead, may yet survive.

8

Conclusion

Ontario Hydro's character and relationship with the Ontario government have been shrouded in ambiguity. An examination of these matters from the creation of the Hydro-Electric Power Commission in 1906 to its formal re-creation as Ontario Hydro in 1973 strikingly confirms the relevance of what Carolyn Tuohy has called the institutionalized ambivalence of Canadian policy and politics. Although attenuated by the revamping, this ambivalence has persisted since but faces new challenges as Hydro struggles with restructuring and privatization in the 1990s. The roots of this ambivalence were apparent from the beginning, stemming from a lack of consensus over the role of the state. This absent consensus was translated into a continuously and deliberately unresolved issue: that of the HEPC's ownership. Was the commission the trustee of a municipal cooperative or a provincial government corporation? There has never been a direct or a frontal answer to this question, not even with the creation of Ontario Hydro. The role of the state has been clothed in the very ambiguity that an absent consensus finds congenial. In the process, the history of the HEPC and Ontario Hydro witnessed repeated demonstrations of what Tuohy terms an 'extraordinary capacity to embody conflicting principles within structures ambiguous enough to allow for ad hoc accommodations over time.'[1] Over time and in varying measure, personalities, legislative statutes, and executive appointments comprised the key elements of this capacity.

Two contemporary analyses of public enterprise have provided useful signposts in tracking the HEPC's and Ontario Hydro's historical development. First, Marsha Chandler's finding that crown corporations fulfil facilitative, redistributive, and nationalistic objectives contributes to an understanding of the varying policy objectives pursued by the commission. In its origin and throughout its history, the utility was expected to be facilitative of economic growth. To the extent it infringed on the property rights of the private sector, it was redistributive. And, at select times, it was nationalistic by explicitly promoting Ontario or Canada over other jurisdictions. Second, Jean Laux and Maureen Molot's view that crown corporations have generational characteristics is applicable for identifying important markers in the HEPC's and Ontario Hydro's development. Although the commission's ambivalent foundations meant that it was not created in the image of the first-generation crown corporations that subsequently appeared, it evolved into one by becoming a sector-wide public ownership monopoly by 1925. By the late 1960s, when it was utilized for industrial strategy and counter-cyclical economic policy, it similarly had evolved to resemble a second-generation crown corporation. When this strategy collapsed in the face of recession and conservation in 1982, Ontario Hydro was forced to begin a not yet completed transformation to a third-generation crown corporation, where business logic replaces social purpose as the operative principle.

The Institutionalized Ambivalence of the HEPC and Ontario Hydro

Electric power policy was critically important to turn-of-the-century Ontario. Technological advances had made the transmission of power over relatively long distances possible, thereby bringing electricity generated from the abundance of water power at Niagara within reach of southwestern Ontario's industrial centres. The advance was economically significant, because it had the potential to relieve the province's dependence on imported coal for generating power. Seeking to exploit the new technology, the owners of the existing electric power monopoly in Toronto created a vertically integrated monopoly of generation, transmission, and distribution in 1902 to serve Toronto with Niagara power. This situation sowed the seeds of subsequent public ownership in Ontario.

The establishment of the vertically integrated Toronto monopoly produced a strong mutual interest among the municipalities of southwestern Ontario. On the one hand, municipal leaders in the small markets and outlying centres feared being cut off from the benefits of Niagara's 'cheap power.' In Toronto, on the other hand, municipal leaders wanted to overcome the domination by the existing private monopoly. When these disparate leaders assembled to discuss their collective plight in 1902, they called for municipal cooperative ownership of a transmission company that would supply municipally-owned distribution commissions as the answer. Liberal premier George Ross addressed the conflicting private and municipal pressures on his government in 1903 by deciding to license the Toronto monopoly's generating station at Niagara and by statutorily permitting municipal cooperative ownership of transmission. The Ross Power Act, however, did not offer the financial assistance necessary to get the cooperative off the ground and did not permit municipalities to compete with private companies on local distribution. The latter meant that the Toronto market, crucial to the viability of the private monopoly and the prospective cooperative, had to be excluded from the municipal scheme. As a result, the government overlooked the facilitative potential of cheap power and certainly did not implement the redistributive policy sought by municipal leaders.

The deficiencies of the Ross Power Act were overcome after the Conservatives took office under James Whitney in 1905. Whitney had won the election in part because, at the behest of Adam Beck, his party had adopted public power as a policy. While simultaneously serving as mayor of London and a Conservative MPP, Beck had been one of the more vocal municipal leaders making demands on the Ross government. The Power Commission Act of 1906, without nationalizing or regulating the vertically integrated monopoly, established the HEPC as a permanent commission of government to transmit power purchased from private generating companies to associated municipally owned local distribution commissions. The fusion of the larger provincial interest with the municipal objective in this manner underscored the government's facilitative objectives for the commission. Moreover, the creation of the HEPC was redistributive because of its intrusion into the private sector: the Power Commission Act established it in co-existence with the private sector on transmission and removed the restrictions on municipalities competing with the private sector on local distribution.

Given that the private sector had already established itself in the industry, the government acted cautiously, choosing not to make a bold statement by creating the HEPC in the image of a first-generation crown corporation or public ownership monopoly. The government instead made a pragmatic, politically rational response to the conflicting pressures it faced, and here began the institutionalized ambivalence of the commission's ownership. Rather than create a new government department with formal ministerial responsibility, the government gave the HEPC the outward character of a crown corporation, a relatively new construct. The collegial decision-making of its three commissioners, although appointed by the government, permitted ostensible autonomy from government. Furthermore, in recognition of the role of the municipalities in leading the government to proceed with the initiative, the HEPC was given two visible characteristics of a municipal cooperative. The municipalities were charged with determining its growth by local referendum and were required to make sinking fund payments to retire its debt, with the latter becoming a basis for claiming equity ownership of the commission. Finally, the corporation and cooperative elements, despite their visibility, were circumscribed by less obvious but powerful departmental controls. Indeed, the HEPC's finances were to be administered like those of a government department and one to two ministers, by statute, were to be among the commissioners, although without any formal ministerial responsibility. Thus, the HEPC was left to resemble a hybrid of instruments – crown corporation, municipal cooperative, and department of government – co-existing with the private sector.

The HEPC evolved into a first-generation crown corporation through the sheer force of the personal leadership of Adam Beck. From his appointment in 1906 as its first chairman until his death in office in 1925, Beck exercised enormous influence over the direction of the commission for two reasons. He sat simultaneously as a minister from 1905 to 1914 and 1923 to 1925, and as an MPP for all but the years 1919–23. Although the presence of a second member of the cabinet sitting as acommissioner, especially between 1906 and 1914, suggested that the HEPC was subject to cabinet control, Beck was able to use his political popularity and that of the commission to dictate policy to his own government. He also enjoyed the extra-parliamentary support of the OMEA, the lobby of the municipally owned distribution commissions. This association was created in 1912 after Beck had created the powerful myth that the HEPC was the trustee of a municipal cooperative rather than a government enterprise. With premiers Whitney, William Hearst, and E.C. Drury reluctant to give the HEPC the autonomy Beck desired, the cooperative claim effectively buttressed the chairman's personal leadership. He used it with great fanfare to heighten the ambivalence of the HEPC's ownership and thereby strengthen the commission's autonomy from government.

Beck succeeded in transforming the HEPC into a first-generation crown corporation through two legendary battles: securing the commission's corporate autonomy from the government, and driving private enterprise out of the industry. In the battle where crown corporation financial practices were at issue, Beck won this autonomy through statutory changes against the better judgment of his own government. Having reduced the HEPC's departmental features primarily to ministerial representation, he had irreversibly altered the weighting of its hybrid composition to a corporation from a department. In the battle over public/private sector co-existence, Beck was uncompromising in his desire to expand public ownership. Rather than begin with nationalizations to rationalize the industry, he used competition assiduously and with great success. By undercutting the rates of private companies and waiting out their leases before proceeding with nationalizations, Beck was able to establish a virtual public ownership monopoly at minimum cost. As a result, the HEPC had become, for all intents and purposes, a first-generation crown corporation by the time of his death. And with his redistributive project completed, the HEPC was left broadly facilitative of economic growth.

With the passing of Beck's personal leadership in 1925, appointments became an instrument for setting the direction of the commission. They were used to institutionalize further the ambivalence of the HEPC's ownership rather than to exercise greater governmental control. Alongside a non-partisan chairman from outside both the government and the HEPC, Premier Howard Ferguson chose to inject political sensitivity into the commission's decision-making. He did so by using the remaining two appointments for elite accommodation. Joining one statutorily authorized cabinet minister, he appointed an OMEA representative, thereby giving the association a status symbolically equal to that of the government. Treating the government and the association as the HEPC's main constituencies, the Ferguson formula for appointments was a masterstroke in political management. Besides reinforcing the statutory confusion over the extent of government ownership, it simultaneously bridged the otherwise disparate provincial and municipal views of the HEPC's ownership and balanced business autonomy and informal political influence. These features of the formula enabled the commission and the government to weather a federal-provincial jurisdictional dispute then occurring over the St Lawrence and Ottawa rivers that threatened the first-generation crown corporation status of the HEPC. Through a nationalistic stance, the commission purchased power from companies wholly within Quebec rather than ceded jurisdiction by buying power from federally licensed companies.

In what ironically turned out to illustrate the Ferguson formula's merit, George Henry, Ferguson's successor, abandoned it in 1930. Henry elevated the minister on the HEPC to the chairmanship. As a result, a critical division of labour which had been established between the commission and the government through a non-partisan chairman was removed. In addition, the other two commissioners left the HEPC even more tied to the government; former Conservative prime minister Arthur Meighen, a representative of no particular constituency, was appointed a commissioner, and the existing OMEA representative emerged publicly as a government partisan. The situation was made worse when the HEPC's chief engineer crossed over a second division of labour with the government by openly attacking Liberal leader Mitch Hepburn for criticizing the commission. Mired in politics and burdened with excess capacity from the Quebec contracts during the depression, the HEPC's inability to sustain itself as a facilitative corporation contributed to the Henry government's defeat in 1934.

After the election, Premier Hepburn reverted to part of the Ferguson formula by appointing a non-partisan chairman, but he diverged from it by appointing two ministers as commissioners. Besides shutting the OMEA out of the HEPC's decision-making, this left the commission subject to cabinet control. The latter permitted a second nationalistic policy and enabled the government to use the power of the legislature to repudiate the HEPC's long-term contracts with Quebec companies. However, with economic recovery creating a power shortage and thereby undermining the wisdom of repudiation, Hepburn returned to the shelter of the Ferguson formula after the 1937 election. By appointing a new non-partisan chairman and again balancing government and OMEA representation, the formula was henceforth elevated to the status of a 'semi-convention.'[2] In this setting, the HEPC abandoned its nationalistic stridency. By negotiating an agreement to develop the Ottawa River, it generally pursued its earlier facilitative crown corporation role, this time to meet the needs of war production.

Following the 1943 election which returned the Conservatives to power, Premier George Drew attempted to infuse the HEPC with policy sensitivity for the government's economic objectives – seen as necessary since the commission was positioned to be facilitative crown corporation at the commanding heights of a booming postwar economy. With the institutionalized ambivalence of the commission, particularly the myth of municipal cooperative ownership, standing in the way of the understanding Drew desired, a tug of war between the government and the OMEA ensued. On the one side, Drew and his successors – Leslie Frost in 1949 and John Robarts in 1961 – resorted to statutory enactments as their weapons in the struggle. Given the political environment of one-party dominance, they were able to act without fear of reprisal in the legislature. On the other side, the OMEA used resolutions at its enormously representative annual conventions to great effect for holding the government at bay. As a result, statutes had increased in importance in relation to appointments, but the principle of balanced government and OMEA representation in the Ferguson formula remained intact.

In the appointments made following the 1943 election, Drew attempted to circumscribe the Ferguson formula by not immediately appointing an OMEA representative. Rather than fill the vacancy, he created an advisory council in 1944 on which he planned to give the association representation. He recanted soon after, however, and appointed an OMEA representative to the HEPC because of pressure from the association. Subsequently, Frost appointed the council in 1951 while respecting the Ferguson formula for the commissioners, only to have the association object to outside advisers for the HEPC. Given this opposition, the government increased the number of commissioners to six in 1955, ostensibly to infuse the commission's decision-making with outside input. The price, however, was increased and disproportionate OMEA representation in the Ferguson formula. The conflict heightened with the establishment of a minister with portfolio for energy in 1959. Here the government won the battle but lost the war as the association successfully resisted the move to make the HEPC formally responsible to the minister. Faced with the resilience of the myth of municipal cooperative ownership, Robarts increased the government's own representation on the commission in 1961 to balance the OMEA in the Ferguson formula. And, after appointing a top civil servant as the senior government representative, the government also ended the last departmental feature of the HEPC in 1962, removing the statutory requirement that a minister sit as a commissioner. With the departure of the last minister in 1963, the tug of war was at a draw.

In the years leading to the re-creation of the HEPC as Ontario Hydro in 1973, the commission's character was noticeably transformed. It moved from a first-generation crown corporation, one passively providing a basic service, to a second-generation crown corporation, one taking a directive role in the economy through counter-cyclical policy. It also moved beyond its merely facilitative role vis-à-vis economic growth to the status of a lead actor in a nationalistic economic strategy. These changes occurred through the HEPC's promotion of the Canadian nuclear industry and Ontario uranium over technology and resources from other jurisdictions. As a result of these developments the political will emerged to overhaul the commission's operation and decision-making. In 1971 Premier Bill Davis initiated a number of studies with the purpose of re-creating the HEPC in the image of a modern crown corporation utilizing contemporary business practices, one by-product of which was an end to the institutionalized ambivalence of its ownership. Given the challenge this presented to the municipal cooperative myth, the cost was a renewed battle with the OMEA. But the latter's reactive, defensive stance would only be able to constrain the initiative, not block the government's overall objective.

While the reform that created Ontario Hydro in 1973 made its corporate features paramount, the institutionalized ambivalence of its predecessor's ownership was only attenuated. The OMEA's opposition to reform ensured that the statutory foundation for the municipal cooperative contention, and most notably the sinking fund payments, survived unamended. However, the symbolic basis for the cooperative contention was severely diminished on three fronts, causing its visibility and vitality to be undermined. First and most strikingly, Ontario Hydro was given a large board of directors. By representing various sectors of the business community and societal interests, not just the government and the OMEA, the new board broke the grip the Ferguson formula had held on the HEPC's decision-making. Secondly, Ontario Hydro, for the first time in its history, was made statutorily responsible to a minister with portfolio for energy. Finally, the OMEA's self-proclaimed role as watch dog over the commission was displaced by external review, although not regulation, of Ontario Hydro by the Ontario Energy Board. As a result, the government institutionalized a new ambivalence. It implemented external public review of a government corporation, one having some municipal cooperative features, alongside a responsible minister with portfolio in a setting where the cabinet still reviewed the corporation's affairs.

The new corporate structure instituted in 1973 improved the working relationship of Ontario Hydro and the Ontario government. This was evident in the first appointments to the new larger board and, for the most part, has remained the case with subsequent appointments and the expansion of the board's size. As for external review, the loopholes in the process that allowed Hydro final authority over rates continue unamended but have served their intended purpose of increased public openness. The circumscribed role of the minister of energy (now environment and energy) in relation to Ontario Hydro persists as well, although much has depended on the personality and the stature of the minister. The municipal cooperative provisions, and particularly the debt retirement scheme, continue to serve as the basis of the myth that Hydro is the trustee of a municipal cooperative, meaning the ambivalence of its ownership has remained a thorny issue. The source of the defence of the municipal cooperative provisions, the Municipal Electrical Association, lacks the political clout of its predecessor, the OMEA, but has compensated with professionalism. The consequence of the revamping may be that Hydro, on balance, has had as much or more autonomy than the HEPC did before 1973.

Since 1973 the assumptions underlying Ontario Hydro's status as a second-generation and a nationalistic crown corporation have collapsed under the weight of conservation and recession. Given the huge excess capacity which resulted, Hydro faced criticism for adopting the attendant policies in the face of such economic uncertainty. In response, in 1982 it abandoned its longstanding practice of planning supply on the basis of anticipated demand, adopting demand management policies instead. This movement toward a third-generation crown corporation, however, proved short-lived. When the economy improved, Hydro proposed a new large-scale nuclear station in 1989, only to cancel its plans when another economic dislocation ensued in 1990. Here began Hydro's second and more concerted attempt to transform itself into a third-generation crown corporation. What has stood in the way of the increased flexibility it desires has been the high visibility of its statutory existence and monopoly status. Although privatization, whether in whole or in part, has been given consideration as a means for Hydro to abandon its past social purpose objectives, to date restructuring initiatives have led the way in the reorientation to business logic.

Political Management through Elite Accommodation

The most enduring innovation to stem from the institutionalized ambivalence of the HEPC was the Ferguson formula for appointments. Prior to its creation, Beck had altered the HEPC's original ambivalence by gaining corporate autonomy for the commission at the expense of departmental controls. In the process, he had increased its ownership ambivalence by strengthening the myth that the HEPC was municipal cooperative. In this setting, Premier Ferguson had little choice but to resort to elite accommodation to manage the contending municipal and provincial ownership claims after Beck's death. Resolving the ownership issue by statute was a political impossibility. Through the Ferguson formula, the OMEA was given an 'institutional foothold'[3] in the HEPC's decision-making symbolically equal to what the government had received from the beginning through statutory representation for the cabinet. As a result, the formula gave visibility and sustenance to the HEPC's institutionalized ambivalence and the municipal cooperative contention in particular. In the process, it served as a mechanism for facilitating accommodations between the government and the OMEA over their conflicting views of the HEPC's ownership and weakened the prospect that the OMEA might mobilize against the government. This had been a tactic which Beck had countenanced and Ferguson had likely considered when the association received its initial representation in 1925.

The Ferguson formula, especially in its original form of a non-partisan chairman from outside the government and the HEPC alongside one government and one OMEA constituency representative, had a number of strengths. For starters, it permitted a division of labour between the government and the chairman, one which precluded cabinet domination because the chairman was not a minister. As for the commissioners, the formula-dictated limit of government representation to a single minister, when by statute there could be two, meant that the government's interests could be heard without the commission being visibly subject to cabinet control. Later, when the HEPC board was increased to six from three commissioners and ministerial representation was removed, the principle of the formula was adapted to an allocation of half of the commissioners for OMEA representatives. While the Municipal Electric Association does not have this predominant influence in Ontario Hydro's affairs, it nonetheless acts as a powerful stakeholder, and municipal hydro leaders have filled not less than two seats on the board of directors and three since 1995.

With respect to the HEPC-staff relationship, the Ferguson formula permitted a second division of labour which further insulated the government from political responsibility for the HEPC's affairs. In modern parlance, this division was between the chairman as chief executive officer and the chief engineer as chief operating officer. As a result, the government could influence the chairman and commissioners, but it was one step removed from directing, or being responsible for, the HEPC's internal affairs. Given the two divisions of labour, the Ferguson formula simultaneously permitted informal influence to be infused into the HEPC's decision-making while visibly respecting the commission's business autonomy. The buffer of these two divisions has been followed with one exception for Ontario Hydro. Although astute political management can be a surrogate for this division between the government and the corporation, if the government selects the president, or the president is too close to the government, both can be open to criticism and the president can be vulnerable if there is a change of government. Now that there is flexibility to appoint either the chair or president as chief executive officer, the government will expose itself to political responsibility for Hydro if it does not resist the temptation to bypass the chair and board for direct relations with a president who is CEO.

The elite accommodation engineered by the Ferguson formula worked in the HEPC case for as long as the range of conflicting interests, in Tuohy's terms, was narrow and all the commissionerswere in 'cross-pressured' positions.[4] The chairman, charged with running the HEPC as a government appointee, assumed the function of mediating the commission's need for business autonomy with the government's desire for political sensitivity or policy sensitivity to its economic objectives. Balancing the view of the OMEA against the government's, a common tactic begun with Beck and carried on by some of his successors, gave the chairman a great deal more autonomy than might otherwise have been the case. The government representative had to express the interests of the government and members of the government caucus while being mindful of the need to distance the government from political responsibility for the HEPC's affairs. Government representatives nonetheless functioned in the shadow of the premier, who as a rule maintained direct relations with the chairman of the commission because of its overwhelming importance to the interests and operation of the government. The OMEA representative, usually a friend of the government, likewise had to express the wishes of the association but had the added burden of being mindful that its claims did not extend too far. If they did, the government could legislate an end to the ambivalence which permitted the myth of municipal cooperative ownership.

Since the establishment of Ontario Hydro's board with broad societal representation, directors have had greater freedom than their commissioner predecessors because the board is not dominated by a small number of stakeholders. The chair remains a mediator of the government's and the corporation's respective needs for political sensitivity and business autonomy. With the diminution of the myth of municipal cooperative ownership, however, the chair does not have as much room for securing Ontario Hydro's autonomy from the government. Indeed, although there has been a formalization of government-corporation relations, most recently through such measures as the memorandum of agreement, governments continue to prefer to manage relations with Ontario Hydro through the more informal means of appointments. This is especially the case for relations with the chair, whose appointment affords the government an important avenue for moral suasion, one without the political responsibility associated with formal levers such as policy directives. A chair selected for his or her policy preference or credentials, be it sustainable development or privatization, would have much more latitude.

The ultimate unwinding of the semi-convention of the Ferguson formula for appointments lay in the fact that it was unable to accommodate the wider range of interests that came to be affected by the HEPC with postwar modernization. In this regard, the HEPC faced the same pressures as other examples of institutionalized ambivalence in Canada, and the Ontario government responded in like manner. In attenuating the ambivalence over ownership in 1973, a new ambivalence was institutionalized among crown corporation decision-making, municipal cooperative-like debt retirement, formal ministerial responsibility, and quasi-regulation by government. This arrangement, which now is under siege with restructuring and possible privatization, has persisted to this day. Although the government could use statutory enactments to end the continuing ownership ambivalence, the action could raise a protracted political and legal conflict that would complicate if not delay its privatization. The political will to initiate privatization, which may have arrived with the outcome of the 1995 election, could likely bring resolution to the ambivalence. As the end of the century approaches, the political culture in Ontario that allowed such ambivalence to thrive appears to be evolving towards a search for certainty and a more limited role for government in the economy.

Appendix
HEPC and Ontario Hydro Leaders, 1906–95

HEPC and Ontario Hydro Chairs

Adam Beck (while minister without portfolio, 1905–14 and 1923–5)	June 1906–August 1925
Charles A. Magrath (while chairman of the International Joint Commission)	September 1925–February 1931
John R. Cooke (while a minister without portfolio)	June 1931–July 1934
T. Stewart Lyon (former editor of the *Globe*)	July 1934–October 1937
Thomas H. Hogg (while chief engineer)	November 1937–February 1947
George H. Challies (acting chairman while minister without portfolio)	February 1947–March 1948
Robert H. Saunders (former mayor of Toronto)	March 1948–January 1955
Richard L. Hearn (interim chairman; former general manager and chief engineer)	January 1955–October 1956

James S. Duncan (former chairman and president, Massey-Ferguson)	November 1956–May 1961
W. Ross Strike (commissioner since 1944; former OMEA president)	June 1961–March 1966
George E. Gathercole (commissioner since 1961; former deputy minister of economics, 1956–62)	April 1966–December 1974
Robert B. Taylor (vice-president, Stelco; member, Task Force Hydro; vice-chairman, Ontario Hydro, 1974)	January 1975–July 1979
Hugh L. Macaulay (private businessman; chair, Ontario Progressive Conservative party, 1971–6)	July 1979–March 1983
Milan Nastich (acting chair while also president)	April 1983–August 1984
Tom Campbell (former senior civil servant, including deputy treasurer)	August 1984–January 1988
Robert C. Franklin (while president of Ontario Hydro; former executive vice-president, CN Rail)	January 1988–May 1991
Marc Eliesen (former deputy minister of energy; chair, Manitoba Hydro)	June 1991–October 1992
Maurice F. Strong (private industrialist; chair, United Nations earth summit, 1992)	December 1992 –
William A. Farlinger (retired chairman, Ernst & Young Canada)	November 1995–

HEPC Commissioners (other than chairs), 1906–74

John S. Hendrie (while minister without portfolio)	June 1906–September 1914

Cecil B. Smith (formerly engineer for Beck's inquiry commission)	June 1906–February 1907
W.K. McNaught (while MPP, 1906–14)	February 1907–February 1919
Isaac B. Lucas (while attorney general, 1914–19)	October 1914–July 1921
Lt.-Col. Dougall Carmichael (while minister without portfolio)	October 1919–June 1923
Fred R. Miller (while member of the Toronto Transit Commission)	July 1921–August 1922
J. George Ramsden (formerly member of Toronto City Council)	January 1923–July 1923
John R. Cooke (while minister without portfolio; continued as chairman)	July 1923–June 1931
C. Alfred Maguire (while president of the OMEA)	September 1925–July 1934
Arthur Meighen (former prime minister of Canada)	June 1931–April 1934
Arthur W. Roebuck (while attorney general)	July 1934–April 1936
Thomas B. McQuesten (while minister of public works and highways)	July 1934–October 1937
William L. Houck (while minister without portfolio)	November 1937–August 1943
J. Albert Smith (while MPP; former OMEA vice-president)	November 1937–August 1943
George H. Challies (while minister without portfolio)	August 1943–May 1955

W. Ross Strike (first appointed when OMEA president; continued as chairman)	June 1944–May 1961
William E. Hamilton (while minister without portfolio)	May 1955–August 1955
Lt.-Col. A.A. Kennedy (first appointed when OMEA president)	May 1955–February 1973
W.K. Warrender (while minister without portfolio)	August 1955–October 1956
T. Ray Connell (while minister without portfolio)	November 1956–May 1958
D.P. Cliff (former OMEA president and its secretary-treasurer to 1965)	November 1956–June 1969
Robert W. Macaulay (while minister without portfolio; later minister of energy resources)	May 1958–October 1963
George E. Gathercole (appointed when deputy minister of economics and development; chairman after 1966)	December 1961–March 1966
William G. Davis (while MPP)	December 1961–February 1962
Robert J. Boyer (while MPP)	November 1962–June 1971
Ian F. McRae (former chairman, Canadian General Electric)	February 1966–September 1970
Dr Douglas J. Flemming (former OMEA president)	October 1969–February 1974
D. Arthur Evans (while MPP; continued thereafter as Ontario Hydro director)	July 1971–February 1974

Lou Danis
(appointed while OMEA vice-president) — September 1971–February 1974

Roger N. Seguin
(Ottawa lawyer) — September 1971–February 1974

J. Dean Muncaster
(appointed while chairman of TFH; continued thereafter as Ontario Hydro director) — February 1973–March 1974

Ontario Hydro Directors (other than chairs and presidents), 1974–95 (outside employment and home town in parentheses)[1]

William Dodge, 1974–80
(former secretary-treasurer, Canadian Labour Congress, Ottawa)

D. Arthur Evans, 1974–5
(MPP; HEPC Commissioner, Bradford)

Andrew Frame, 1974–5
(former OMEA president; member of Task Force Hydro, Burlington)

Douglas G. Hugill, 1974–6
(former OMEA president; tax consultant, Sault Ste Marie)

Allen T. Lambert, 1974–80
(chairman and CEO, Toronto-Dominion Bank, Toronto)

J. Conrad Lavigne, 1974–81
(president, Mid Canada Television Systems, Timmins)

Philip B. Lind, 1974–81
(vice-president and secretary, Rogers Cable Systems, Toronto)

J. Dean Muncaster, 1974–80
(president and CEO, Canadian Tire Corporation, Toronto)

A.C. Jean Pigott, 1974–76
(president and CEO, Morrison-Lamothe Foods, Ottawa)

Robert J. Uffen, 1974–80
(dean of applied science, Queen's University, Kingston)

Robert H. Hay, 1975–8
(former OMEA president, Kingston)

William A. Stewart, 1976–85
(former minister of agriculture and food, Denfield)

Robert M. Schmon, 1976–8
(president and CEO, Ontario Paper Company, Niagara-on-the-Lake)

Arthur J. Bowker, 1977–85
(former OMEA president; senior research officer, National Research Council, Ottawa)

1 Information culled primarily from Ontario Hydro annual reports.

Sister Mary, 1977–81
(executive director, St Michael's Hospital, Toronto)

Alan B. Cousins, 1979–88
(former OMEA president; president, Ideal Stamping, Wallaceburg)

A. Ephraim Diamond, 1979–86
(former chairman and CEO, Cadillac Fairview, North York)

J.A. Gordon Bell, 1981–92
(president and CEO, Bank of Nova Scotia, Thornhill)

Albert G. Hearn, 1981–92
(former vice-president, Service Employees International Union, Agincourt)

Dr O. John C. Runnalls, 1981–95
(professor emeritus, Nuclear Engineering and Energy Studies, University of Toronto, Toronto)

Leonard N. Savoie, 1981–7
(president and CEO, Algoma Central Railway, Sault Ste Marie)

John B. Cronyn, 1982–8
(director, John Labatt; former chairman, Committee on Government Productivity, London)

John W. Erickson, 1982–8
(barrister and solicitor, Thunder Bay)

Isobel Harper, 1982–8
(president, BDI Enterprises, Toronto)

Richard E. Cavanagh, 1985–91
(former OMEA president; chairman, Scarborough PUC, Scarborough)

F. Tom Cowan, 1985–91
(Chimo Farms, Mount Bridges)

Alexander J. MacIntosh, 1986–91
(partner, Blake, Cassels & Graydon, Toronto)

James S. Hinds, 1987–
(barrister and solicitor, Hinds and Sinclair, Sudbury)

John E. Hood, 1988–91
(vice-chairman, Stelco, Toronto)

Dr Geraldine A. Kenny-Wallace, 1988–91
(president, McMaster University, Toronto)

Gaston Malette, 1988–91
(chairman and CEO, Waferboard Corporation, Timmins)

John E. Kennedy, 1990–3
(vice-president, Midland Walwyn Capital, Thunder Bay)

Dr Mohan Mathur, 1990–
(dean of engineering science, University of Western Ontario, London)

Joseph R. O'Brien, 1990–3
(former MEA chair; president, St Catharines Paving, St Catharines)

Dr Mary Jane Ashley, 1991–4
(chair, Department of Preventive Medicine, University of Toronto, Toronto)

David B. Brooks, 1991–4
(associate director, International Development Research Institute, Ottawa)

Michael Cassidy, 1991–
(former Ontario NDP leader; president, Ginger Group Consultants, Ottawa)

Adèle M. Hurley, 1991–
(president, Hurley and Associates [Environmental Consultants], Toronto)

Elmer McVey, 1991–
(former chair, Sudbury Hydro, Sudbury)

Anne A. Noonan, 1991–
(native affairs consultant, Anne A. Noonan & Associates, Ottawa)

Kealey C. Cummings, 1992–
(former secretary-treasurer, Canadian Union of Public Employees, Ottawa)

George Davies, 1992
(deputy minister of energy, Toronto – non-voting director)

D. Murray Wallace, 1992–4
(president and CEO, AVCO Financial Services, London)

Richard Dicerni, 1993–5
(deputy minister of environment and energy, Toronto – non-voting director)

Nuala Beck, 1993–
(president, Nuala Beck & Associates, Mississauga)

Eleanor Clitheroe, 1993–
(senior vice-president and chief financial officer, Ontario Hydro, Toronto)

Bill Etherington, 1993–5
(president and CEO, IBM Canada, Thornhill)

Dona Harvey, 1993–
(journalist, Kitchener)

David Kerr, 1993–
(president and CEO, Noranda Inc., Toronto)

Doug McCaig, 1993–
(former MEA chair, Fort Frances)

Andrew Sarlos, 1993–
(president, Andrew Sarlos and Associates, Toronto)

Arthur Sawchuk, 1993–
(president and CEO, Du Pont Canada, Toronto)

John Murphy, 1994–
(president, Power Workers' Union, Toronto)

Carl Anderson, 1995–
(chair, North York Hydro; former MEA chair)

Donald Fullerton, 1995–
(chair and CEO, Canadian Imperial Bank of Commerce, Toronto)

Jim MacNeill, 1995–
(president, MacNeill & Associates; chair, Earth Council Foundation [Washington], Toronto)

Robert Schad, 1995–
(president, Husky Injection Molding Systems, Toronto)

Government Representatives on the HEPC and Ontario Hydro (tenure with the HEPC and Hydro in parentheses)

Adam Beck (1906–25)
Minister without portfolio, 1905–14 and 1923–5, MPP, 1902–19 and 1923–5

John S. Hendrie (1906–14)
Minister without portfolio, 1905–14

W. K. McNaught (1907–19)
MPP, 1906–14

Isaac B. Lucas (1914–21)
Attorney general, 1914–19

Lt.-Col. Dougall Carmichael (1919–23)
Minister without portfolio

John R. Cooke (1923–34)
Minister without portfolio

Arthur W. Roebuck (1934–7)
Attorney general

Thomas B. McQuesten (1934–7)
Minister of public works and highways

William L. Houck (1937–43)
Minister without portfolio

J. Albert Smith (1937–43)
MPP

George H. Challies (1943–55)
Minister without portfolio

William E. Hamilton (1955)
Minister without portfolio

William K. Warrender (1955–6)
Minister without portfolio

T. Ray Connell (1956–8)
Minister without portfolio

Robert W. Macaulay (1958–63)
Minister without portfolio, 1958–9, Minister of energy resources, 1959–63

George E. Gathercole (1961–74)
Deputy minister of economics, 1956–62

William G. Davis (1961–2)
MPP

Robert J. Boyer (1962–71)
MPP

D. Arthur Evans (1971–5)
MPP

George Davies (1992)
Deputy minister of energy – non-voting

Richard Dicerni, (1993–5)
Deputy minister of environment and energy – non-voting

OMEA/MEA Representatives on the HEPC and Ontario Hydro (tenure with the HEPC and Ontario Hydro in parentheses)

C. Alfred Maguire (1925–34)
OMEA president, 1923–35

J. Albert Smith (1937–43)
OMEA vice-president, 1936–7

W. Ross Strike (1944–66)
OMEA president, 1944–6

Lt.-Col. A.A. Kennedy (1955–73)
OMEA president, 1953–6

D.P. Bud Cliff (1956–69)
OMEA president, 1950–2, secretary-treasurer, 1953–65

Dr Douglas J. Flemming (1969–74)
OMEA president, 1967–8

Lou Danis (1971–4)
OMEA vice-president, 1971

Andrew Frame (1974–5)
OMEA president, 1971–2

Dr Robert H. Hay (1974–9)
OMEA president, 1965–6

Douglas G. Hugill (1975–7)
OMEA president, 1970–1

Arthur J. Bowker (1977–85)
OMEA president, 1974–5

Alan B. Cousins (1979–88)
OMEA president, 1975–6

Richard E. Cavanagh (1985–91)
OMEA president, 1982–3

Joseph R. O'Brien (1990–3)
MEA chair, 1986–7

Elmer McVey (1991–)
Former chair, Sudbury Hydro

Doug McCaig (1993–)
MEA chair, 1992–3

Carl Anderson, (1995–)
Chair, North York Hydro; MEA chair, 1988–90

Ministerial Supervision of the HEPC and Ontario Hydro[2]
(party affiliation in parentheses)

Adam Beck (Conservative)	
Minister without portfolio	7 June 1906–2 October 1914
John S. Hendrie (Conservative)	
Minister without portfolio	7 June 1906–2 October 1914
Isaac B. Lucas (Conservative)	
Treasurer	2 October 1914–22 December 1914
Attorney General	22 December 1914–14 November 1919
Lt.-Col. Dougall Carmichael (United Farmers)	
Minister without portfolio	14 November 1919–16 July 1923
Adam Beck (Conservative)	
Minister without portfolio	16 July 1923–15 August 1925
John R. Cooke (Conservative)	
Minister without portfolio	16 July 1923–10 July 1934
Arthur W. Roebuck (Liberal)	
Attorney General	10 July 1934–14 April 1937

2 Information drawn from *Legislators and Legislatures of Ontario: A Reference Guide*.

Thomas B. McQuesten (Liberal)	
Minister of highways	10 July 1934–1 November 1937
Minister of public works	
William L. Houck (Liberal)	
Minister without portfolio	1 November 1937–17 August 1943
George H. Challies (Conservative)	
Minister without portfolio	17 August 1937–2 May 1955
William E. Hamilton (Conservative)	
Minister without portfolio	2 May 1955–17 August 1955
William K. Warrender (Conservative)	
Minister without portfolio	17 August 1955–1 November 1956
T. Ray Connell (Conservative)	
Minister without portfolio	1 November 1956–28 April 1958
Minister of reform institutions	28 April 1958–26 May 1958
Robert W. Macaulay (Conservative)	
Minister without portfolio	26 May 1958–5 May 1959
Minister of energy resources	5 May 1959–16 October 1963
Minister of economics and development	8 November 1961–16 October 1963
J.R. Jack Simonett (Conservative)	
Minister of energy resources	16 October 1963–25 March 1964
Minister of energy and resources management	25 March 1964–5 June 1969
George A. Kerr (Conservative)	
Minister of energy and resources management	5 June 1969–23 July 1971
Minister of environment	23 July 1971–2 February 1972
James A.C. Auld (Conservative)	
Minister of environment	2 February 1972–4 July 1973
W. Darcy McKeough (Conservative)	
Minister of energy	4 July 1973–18 June 1975
Dennis R. Timbrell (Conservative)	
Minister of energy	18 June 1975–3 February 1977
James A. Taylor (Conservative)	
Minister of energy	3 February 1977–21 January 1978

Rueben C. Baetz (Conservative) Minister of energy	21 January 1978–18 August 1978
James A.C. Auld (Conservative) Minister of energy	18 August 1978–30 August 1979
Robert S. Welch (Conservative) Minister of energy Deputy premier	30 August 1979–6 July 1983
Philip Andrewes (Conservative) Minister of energy	6 July 1983–8 February 1985
George L. Ashe (Conservative) Minister of energy	8 February 1985–17 May 1985
Michael D. Harris (Conservative) Minister of energy	17 May 1985–26 June 1985
Vincent G. Kerrio (Liberal) Minister of energy	26 June 1985–29 September 1987
Robert C. Wong (Liberal) Minister of energy	29 September 1987–2 August 1989
Lyn McLeod (Liberal) Minister of energy	2 August 1989–1 October 1990
Jenny Carter (New Democrat) Minister of energy	1 October 1990–31 July 1991
William A. Ferguson (New Democrat) Minister of energy	31 July 1991–13 February 1992
Brian A. Charlton (New Democrat) Minister of energy Minister of financial institutions	13 February 1992–3 February 1993
C.J. Bud Wildman (New Democrat) Minister of energy Minister of environment	3 February 1993–26 June 1995

HEPC and Ontario Hydro Senior Administrative Officers

P.W. Southam
Chief engineer, 1906–12

Fredrick A. Gaby
Chief engineer, 1912–34

Richard T. Jeffery,
Chief engineer, municipal relations and rural power, 1934–7

Thomas H. Hogg
Chief engineer, hydraulic and operation, 1934–7

Thomas H. Hogg
Chief engineer, 1937–47 (while chairman)

Richard L. Hearn
General manager and chief engineer, 1947–55

A.W. Manby
General manager, 1955–9

J.M. Hambly
General manager, 1960–70

Douglas J. Gordon
General manager, 1970–4, president, 1974–80

Milan Nastich
President, 1980–5 (while acting chairman, 1983–4)

Robert C. Franklin
President, 1986–91 (president and CEO, 1989–91; chairperson, 1988–91)

Alan R. Holt
President, 1991–2

O. Allan Kupcis
President, 1992– (president and CEO, 1995–)

Notes

1: Introduction

1 Tuohy, *Policy and Politics in Canada.*
2 Ibid., 5.
3 Heard, *Canadian Constitutional Conventions*, 145–8.
4 Black and Cairns, 'Different Perspective on Canadian Federalism.' See also Young, Faucher, and Blais, 'Province-Building: A Critique.'
5 For the Ontario case specifically, see H.V. Nelles's *Politics of Development.*
6 See, for example, Chandler and Chandler, *Public Policy and Provincial Politics*, 12 and 253.
7 Chandler, 'The Politics of Public Enterprise,' 209–12.
8 Laux and Molot, *State Capitalism*, 4, 5, 20–1, 24, 35, and 173–4. The concept of first and second generations of crown corporations as applied to Canada is borrowed from Holland, 'Europe's New Public Enterprises.' As an addition to the first and second generations, Laux and Molot devised the third.

2: Creating the HEPC, 1902–6

1 On this perspective, see Trebilcock, Prichard, Hartle, and Dewees, *Choice of Governing Instrument.*
2 Plewman, *Beck*, 28; Nelles, *Politics of Development*, 216–17. Reference to secondary sources will be made only for instances of direct quotations, notable observations, and findings that differ from my own. For a detailed tracking of these sources, see Freeman, 'Ontario Hydro and Its Government.'
3 Armstrong and Nelles, *Monopoly's Moment*, 141–2, 146, 154, and 160.
4 The Municipal Amendment Act, SO 1899 c. 26 s. 35.
5 Nelles, *Politics of Development*, 239–41.

6 Plewman, *Beck*, 36; Humphries, *Whitney*, 85–6.
7 Quoted in Plewman, *Beck*, 39.
8 An Act to provide for the Construction and Distribution of Municipal Power Works and the Transmission, Distribution and Supply of Electrical and other Power and Energy, SO 1903 c. 25.
9 The cabinet's only role was to determine the level of compensation for the chief justice's services. See Mavor, *Niagara in Politics*, 35; SO 1903 c. 25 s. 51.
10 SO 1903 c. 25 s. 1, 2, 52, 11, 9, and 10.
11 Ibid., ss. 12, 49, 17, 22, 21, 23, and 24.
12 Ibid., ss. 41, 42, and 26. The bond scheme, minus the roles for the trustee and chief justice and the special bonds, was first outlined by Alderman Spence at the Berlin meeting in 1902. See Bolton, *Expensive Experiment*, 19–20.
13 Plewman, *Beck*, 40 and 239.
14 Nelles, *Politics of Development*, 247; Humphries, *Whitney*, 96.
15 Nelles, *Politics of Development*, 247.
16 *Canadian Annual Review, 1905*, 286.
17 Hodgetts, *Arm's Length to Hands-On*, 10 and 83; Humphries, *Whitney*, 96.
18 *Canadian Annual Review, 1905*, 288–9.
19 For the details of the HECI mandate, see Ontario, *Hydro-Electric Power Commission: First Report* (hereafter the Beck Report). The commission is referred to here as the HECI (which is the title used by Nelles), even though 'inquiry' is not in its official name. I give it this name to distinguish it from the permanent commission established in 1906 which adopted the inquiry's official name.
20 Denison, *People's Power*, 47.
21 Beck Report, 5 and 3.
22 Ontario Power Commission, *Official Report* (hereafter Snider Report); Beck Report; Nelles, *Politics of Development*, 259.
23 Plewman, *Beck*, 46–7.
24 The original scheme of the municipal movement was to avoid the capital cost of generation, but the capital costs presented in the report are based on 'development, transmission and distribution.' Snider Report, 10. Nelles claims the 'Commissioners decided for the time being that they could not afford to build a generating station.' Nelles, *Politics of Development*, 263.
25 Snider Report, 20, 9–10, and 13.
26 Ibid., 16 and 27–8.
27 Ibid., 20–2 and 31–2; Nelles, *Politics of Development*, 263. In correspondence with the author, Nelles writes that his reference to regulation should have been to the 'proposed cooperative enterprise itself.'
28 Nelles, *Politics of Development*, 263. In correspondence with the author, Nelles writes that he understood the word 'interfere' to mean that the 'Conmee Act

would have to be modified rather than that the project would have to be abandoned.' Although this conclusion is a reasonable one, it does not necessarily follow that a united stance on the need to proceed meant that the commissioners were not split on the issue of repeal. This makes the word 'interfere' a simple factual statement, lacking a definitive recommendation on how to proceed. Snider's property in Waterloo is noted in Ontario, *Report on History and General Relations*, 13, and his investment in Algoma is noted in Plewman, *Adam Beck*, 239.

29 Snider Report, 18.

30 Mavor, *Niagara in Politics*, 44.

31 Nelles, *Politics of Development*, 264. As another example of the mythology, one with a different conclusion than Nelles's, Dewar writes that the Beck report 'did not favour public ownership of generation, but it did recommend that the province build the transmission facilities.' Dewar, 'The Origins of Ontario Hydro,' 208.

32 Beck Report, 7.

33 Ibid., 10 and 11. Whitney issued orders-in-council requesting the figures. See Humphries, *Whitney*, 152.

34 Beck Report, 8, 10, A8, and A14.

35 Plewman, *Beck*, 48.

36 Ibid., 48.

37 Plewman also states that Allan Dymond, later chief law clerk of Ontario, drafted the act, but in light of his statement regarding Meredith he probably means to suggest that Dymond was the law clerk when it was drafted. Ibid., 49 and 448.

38 Meredith, as cited in Humphries, *Whitney*, 154, and in Nelles, *Politics of Development*, 265; Ontario, *Bills, 1906*, n. 243; An Act to provide for the Transmission of Electrical Power to the Municipalities, SO 1906 c. 15; *Globe*, 9 and 10 May 1906.

39 Ross's objections to the bill were the effect on credit, the politicization of the HEPC's affairs through cabinet ministers serving as commissioners, the provision for provincial rather than municipal debt, provincial debt for only one region, the override of the Conmee Clause, and the inclusion of expropriation powers. Ontario, *Report on History and General Relations*, 47; *Globe*, 10 and 11 May 1906.

40 On this conclusion, Nelles is in agreement. In correspondence he writes: 'I agree with you entirely that a process of hybridization was involved, a process which created an organization that was between things, between the province and the municipalities, and thus effectively an agency on its own. Whether that was understood at the time I am not certain. I think not. Rather I expect people thought they were squaring circles.'

41 Underlying the inclusion of this supervisory role, in Brady's view, was the desire

that the 'government and commission ... were to act in unison; and that the major decisions of the commission, especially with regard to expenditures were to come regularly before the government.' Brady, 'Ontario Hydro-Electric Power Commission,' 331–2; SO 1906 c. 15 s. 1.

42 SO 1906 c. 15 ss. 1, 2, and 3; Ontario, *Bills, 1906*, n. 243 s. 1; *Globe*, 10 May 1906, 1 and 8.
43 SO 1906 c. 15 ss. 11, 12, 13, 21, and 22.
44 Ibid., ss. 18, 17, 4, and 5.
45 Ibid., ss. 14, 15, and 16.
46 Ibid., ss. 6 and 7.
47 Ibid., ss. 10 and 8.
48 The bill, as amended, deleted references to natural gas and added that electricity regulation only affected those bodies 'receiving power from the Commission.' Ontario, *Bills, 1906*, n. 243 s. 19; State of New York, *Report on the Conservation of Water*, *1912*; *Globe*, 10 May 1906, 1; SO 1906 c. 15 s. 19.
49 SO 1906 c. 15 ss. 7 and 8.

3: The Beck Era, 1906–25

1 Quoted in Nelles, *Politics of Development*, 266–7.
2 Dales, *Hydro-Electricity and Industrialization*.
3 The claim that the HEPC was only the 'trustee' of a municipal cooperative did not begin with this bylaw. The Toronto bylaw, if in fact the same as the others, did not speak of the HEPC in these terms. City of Toronto, *Minutes*, 1907, Bylaw 4834, Appendix B, 6–7.
4 SO 1907 c. 19 ss. 8(b), 8(c), and 15; Mavor, *Niagara in Politics*, 92. The publication of Mavor's book was financially assisted by the National Electric Light Association, a private enterprise lobby in the United States. Plewman, *Beck*, 436.
5 Quoted in both Nelles, *Politics of Development*, 273–4, and Humphries, *Whitney*, 165.
6 Mavor, *Niagara in Politics*, 87–90; *Canadian Annual Review, 1907*, 517–18.
7 SO 1909 c. 19, Schedule A, ss. 12, 2(a), and 2(d); OMEA, *Submission to the Premier*.
8 Quoted in Nelles, *Politics of Development*, 288.
9 The 1908 statute permitted the mayors of municipalities to sign HEPC contracts without returning to the electors to approve the discrepancy. SO 1908 c. 22. The 1909 statute also included the contract let for the construction of a transmission line in direct competition with the TPC. SO 1909 c. 19 ss. 5, 8, and 9. To overcome such problems in the future, the 1908 statute had introduced a

standard municipal-HEPC contract for all future bylaw referendums. SO 1908 c. 22 Appendix B. The problem which underlay the controversy was the municipal cooperative dimension of the HEPC's operation. It left the onus for the wording of the bylaw with the municipalities, terms which they had set collectively in the summer of 1906. The new standard contract, however, was written in a way that suggested the HEPC was a trustee of a municipal cooperative, although no statutory basis for a trustee relationship existed. For an early critical assessment of the argument that the HEPC was a trustee of the municipalities, see Ontario, *Hydro-Electric Inquiry Commission: General Report* (hereafter Gregory Report), 29–32.

10 SO 1910 c. 16 s. 4.

11 According to Armstrong, it was at this point in Canadian history that 'disallowance (and even the threat of disallowance) largely ceased to be a means by which the federal government could exercise discipline over the provinces.' Armstrong, *Politics of Federalism*, 65.

12 The bill, No. 102, was entitled An Act to Amend the Power Commission Act and the Ontario Railway and Municipal Board Act. Ontario, *Journals, 1911*, xix and 81.

13 Plewman, *Beck*, 82–4; Nelles, *Politics of Development*, 400–1.

14 Prang, *Rowell*, 107; Plewman, *Beck*, 83–4; Denison, *People's Power*, 99.

15 Nelles, *Politics of Development*, 303; Plewman, *Beck*, 99–100.

16 SO 1912 c. 14 ss. 2, 9, and 15; Prang, *Rowell*, 107–8.

17 The Ferris Committee was responding to calls for the New York State Conservation Commission to own and operate a transmission system. New York State, *Report on the Conservation of Water, 1912* (hereafter Ferris Report, 1912); and New York State, *Report on the Conservation of Water, 1913* (hereafter Ferris Report, 1913). According to Plewman, Beck felt the Ferris Committee was biased against public ownership; he therefore did everything in his power to stymie its gathering of information. Beck, however, gave testimony to the committee in New York City in 1911, and again in Ottawa in 1912. It was only after this second appearance that Beck no longer cooperated with the committee. The minutes are replete with Senator Ferris's unsuccessful appeals for Beck to cooperate. Under these circumstances, it is not surprising that the majority report in 1913 rested almost exclusively on the testimony of Reginald Pelham Bolton, a New York engineer who gleaned his information from interviews and documents he received from a dismissed and disgruntled HEPC employee. Bolton's testimony accounts for 318 of the 699 pages of minutes of the 1913 report. A minority report, however, claimed that Bolton had 'prejudged the situation,' had gone to Ontario in 'search of failure,' and had relied on the hearsay evidence of a per-

son who had refused to testify before the committee. Plewman, *Beck*, 107–8; Ferris Report, 1913, xx.

18 Bolton, *Expensive Experiment*, 79, 278, and 172; Plewman, *Beck*, 113; Denison, *People's Power*, 103–4.

19 Mavor, *Niagara in Politics*, 110.

20 The issue of the division of labour in the public sector surfaced after Beck had used the HEPC's statutory power to set municipal rates on 13 November 1913 and Toronto Hydro refused the order. This rate problem remained unresolved until the world war led to reductions greater than those originally requested. SO 1909 c. 19 schedule A; SO 1912 c. 14 s. 9. Although Plewman writes that Black was the HEPC's nominee to Toronto Hydro, no provision for such an appointment existed in the Public Utilities Act or the Power Commission Act, although such a provision was added to the latter in 1915. Plewman, *Beck*, 127 and 136; RSO 1914 c. 204 s. 36; SO 1915 c. 19. s. 15.

21 The Hydro-Electric Railway Act, SO 1914 c. 31; SO 1914 c. 16 s. 4; Plewman, *Beck*, 134 and 175–6. In 1915 the Power Commission Act was amended again to include the commissioners under the same exemption from the Legislative Assembly Act that the chairman had enjoyed since 1912 upon receiving a salary. This amendment was made retroactive to 1 November 1914. SO 1915 c. 19 s. 2. Beck received $6,000 in appropriations (since 1912), $6,000 from the municipal hydro commissions, $6,000 as a cabinet minister, and $1,400 as a sessional indemnity for a total of $19,400. Hendrie received $4,500 from the municipal hydros, $6,000 as a cabinet minister, and the $1,400 indemnity for a total of $11,900. McNaught received the $4,500 and the indemnity for a total of $5,900.

22 Nelles, *Politics of Development*, 401.

23 SO 1915 c. 19 s. 15.

24 SO 1916 c. 37 s. 9; Oliver, *Ferguson*, 81.

25 The Central Ontario Power Act, SO 1916 c. 18; Oliver, *Ferguson*, 64 and 81; Oliver, 'Sir William Hearst,' 27; Plewman, *Beck*, 193–4.

26 Ontario, 'Public Accounts, 1914–15,' 555. The tale is recounted in Hodgetts, *From Arm's Length to Hands-On*, 149–50.

27 Ontario, 'Report of the Committee on Public Accounts, 1916,' 116–18. The controversy was heightened by the release of a series of articles critical of the HEPC which appeared in the *Financial Post* in 1916 by James Mavor. The articles were reprinted as Mavor, *Public Ownership and the Hydro-Electric Power Commission*.

28 Ontario, 'Report of the Committee on Public Accounts, 1916,' 125–30 and 62.

29 Ibid., 12–13, 17, 19, 60–1, 29–30, and 64–5.

30 Clarkson's son and successor, chartered accountant Geoffrey T. Clarkson, revealed the substance of the recommendation to McGarry in testimony before

the Gregory Commission. Gregory Report, 55. The Clarkson audit was undertaken only to ensure that proper documentation existed for all HEPC expenditures to 1916. Clarkson was not asked to determine whether the expenditures had the legislature's approval.

31 Another 1916 amendment provided authorization for the deferment of sinking fund payments. Although the practice had begun with the construction of the transmission line in 1909, it had never been formalized in the act, a situation to which the provincial auditor took exception. The amendment permitted the cabinet to relieve a municipality from the first five of thirty years of payments, with the total due over the remaining twenty-five. The provision was amended in 1917, however, to permit the HEPC rather than the cabinet to authorize the relief, with the payments made over a subsequent thirty years for a total of thirty-five years. SO 1916 c. 19 s. 12; SO 1917 c. 20 s. 13. Neither amendment was made retroactive to 1909 to legitimate the deferments made prior to this statutory authorization.

32 These 1916 amendments were made retroactive to 31 October 1910 to legitimize the HEPC's past application of revenue to expenditures. The commission was also permitted to purchase and sell supplies out of its own funds and charge and collect from municipal hydro commissions for consulting and contracting work. Both had been non-statutory practices to which the provincial auditor objected. SO 1916 c. 19 ss. 6, 8, and 4; Hodgetts, *Arm's Length to Hands-On*, 150.

33 SO 1916 c. 19 s. 4. The provision did not, however, relieve the HEPC of s. 14 of RSO 1914 c. 39 which stated that funds raised for the HEPC were to be audited and accounted for in the same manner as other public revenue and accounts.

34 SO 1917 c. 20 s. 2; SO 1918 c. 14 s. 3; SO 1927 c. 17 s. 8; Gregory Report, 53 and 187. Following the 1916 requirement for an annual report, the HEPC began to list the municipal sinking fund contributions in 1917. After this became a requirement in 1918 the commission began to list the contributions as equity in 1919.

35 The act also specified that a separate account was to be kept for the project, and it spelled out that all costs were to be defrayed out of appropriations, with expenditures not to exceed appropriations in any one year. The Ontario Niagara Development Act, SO 1916 c. 20. ss. 4 and 5.

36 Nelles, *Politics of Development*, 404.

37 SO 1917 c. 21 ss. 9 and 6; Plewman, *Beck*, 200.

38 SO 1916 c. 20 s. 7.

39 Plewman, *Beck*, 206–7; Armstrong, *Politics of Federalism*, 77–8 and 77n.

40 SO 1917 c. 20 ss. 3 and 5; Plewman, *Beck*, 201–2.

41 In addition, the inspector was permitted to recommend to cabinet that limits on quantities of water utilized be imposed. The limits could be imposed whether or not rights had already been conferred. The act also provided that compensation

could be applied for and that the inspector's orders could be appealed to the cabinet. SO 1916 c. 21 ss. 4, 5, 9, and 10; Plewman, *Beck*, 202–3.

42 The Water Powers Regulation Act, SO 1917 c. 22.

43 Plewman, *Beck*, 208–9; Armstrong, *Politics of Federalism*, 78–9; Nelles, *Politics of Development*, 365.

44 Plewman, *Beck*, 221–2.

45 SO 1919 c. 45 s. 2.

46 Although Whitney before Hearst and Ferguson after him had similar conflicts with Beck, Oliver writes that neither 'permitted relations to deteriorate to the extent that they threatened the existence of the government.' For this reason, Oliver concludes that Beck's reckless conduct correlated with the 'absence of a strong Premier.' Oliver, 'Sir William Hearst,' 36–7. See also Tennyson, 'Beck and the Election of 1919.'

47 There is some controversy over who turned down whom on the offer of the premiership, but in sum, Beck wanted to have complete freedom from UFO party policy on HEPC matters. See Johnson, *Drury*, 60–1.

48 Denison, *People's Power*, 144.

49 Nelles, *Politics of Development*, 413–16.

50 Sutherland's terms of reference also included the outlook for radials and whether they could be self-supporting, their effect on Canadian National Railways (just created), their effect on national, provincial, and municipal debt, and whether they were a public necessity and within the objects of the HEPC. Sutherland was a former Liberal MP and speaker of the House of Commons. He was also one of the judges appointed by Beck under the Water Powers Regulation Act, 1917, to report on the TPC. Ontario, *Report on Hydro-Electric Radials, 1921* (hereafter Sutherland Report); Johnson, *Drury*, 108–13.

51 The Toronto Power and Railway Purchase Act, SO 1921 c. 23; Plewman, *Beck*, 259–60.

52 The majority report considered existing radials in Canada and the United States to be in a 'precarious and unsatisfactory' position with a 'dubious and discouraging' future. In addition, it considered radial competition with the newly formed Canadian National Railways to be 'unwise and economically unsound.' And it condemned the practice of guaranteeing municipal radial bonds and suggested that the apprehension over the size of government debt in Canada was a cogent reason not to proceed with radials. The minority report recommended the government endorse the principle of publicly owned radials and instruct the HEPC to proceed with its plans. Sutherland Report, 3–5, 182, and 211. The delay in the release of Beck's counter report occurred because the Sutherland report was only officially published in December 1921. HEPC, *Statement Re: Majority Report of the Sutherland Commission.*

53 The Murray report was commissioned by the National Electric Light Association, a pro-private enterprise lobby in the United States. The HEPC's refutation maintained that the commission was created after a consideration of all factors, including American-style regulation which it felt to be ineffective in 'numerous instances.' In general, the HEPC argued that the real purpose of the Murray report was 'to undermine, if not destroy, public confidence in public ownership' by discrediting the commission. It also argued that the Murray report failed to explain fully the HEPC's organization as a trustee for a municipal cooperative, preferring instead to attach the stigma of government ownership. Murray, *Government and Private Electric Utilities in Canada and the United States*, 1, 3, and 16; HEPC, *Refutation of 'Government and Private Electric Utilities.'*

54 Although seven of the Gregory Commission's eight terms of reference concerned the Queenston project exclusively, the eighth gave it the power to examine 'any other power development ..., and generally all matters of expenditure and administration.' Plewman, *Beck*, 297–8; Gregory Report, 1. Gregory was a Toronto lawyer and a Liberal who had been a director of the *Weekly Sun* farmers' newspaper and was known to be a friend of the farmers' cause. Johnson, *Drury*, 107.

55 The Municipal Electric Railways Act, SO 1922 c. 69. As part of the compromise, Beck agreed to terms with the TPC for the purchase of its suburban railways, although the generation and transmission properties still needed to be negotiated. The Toronto Suburban Railway Company Act, SO 1922 c. 35. It should be noted that the transfer of property rights for Beck's ambitious plans for east and west radial entrances to Toronto was defeated in a referendum in Toronto on 1 January 1923. In addition, Hamilton and lesser municipalities crucial to the Toronto–St Catharines radial rejected the scheme on this same date. This led Drury to authorize only the Toronto–Oakville section, which enjoyed municipal support, although the resolution had not been to Beck's liking. Plewman, *Beck*, 329–31. With the HEPC having already spent monies provided by municipalities the shortened line did not reach, St Catharines launched a court action for their recovery. JCPC, *St Catharines* v. *Hydro-Electric Power Commission*.

56 Brady, 'Hydro-Electric Power Commission,' 337.

57 Johnson, *Drury*, 204; Plewman, *Beck*, 338–9.

58 Ibid., 335 and 345.

59 Plewman offers another reading of the order-in-council. He claims the amendment was intended to delete the eighth term of reference, of which the rider was considered a part, because its open-ended wording had broadened the Gregory Commission into a full-scale inquiry of the HEPC's operation rather than just the Queenston project. Thus, Plewman argues that the commission defied the government's order. This conclusion, however, is not substantiated from an

examination of the order-in-council. The answer to this misunderstanding may lie in Plewman's own analysis. He suggests Ferguson let the commission know he was bluffing, which fits Oliver's view that Ferguson knew the Gregory report would 'strengthen his own hand' when dealing with Beck and the HEPC. Ontario, Order-in-Council, 12 September 1923; Plewman, *Beck*, 331 and 352–3; Oliver, *Ferguson*, 152; Gregory Report, 1–2.

60 Gregory Report, 235–6, 27, and 225.

61 Ibid., 29–32 and 68–9.

62 Ibid., 229, 219–20, and 222–3.

63 Ibid., 147 and 219; HEPC, *Errors and Misrepresentations made by the Hydro-Electric Inquiry Commission.*

64 SO 1924 c. 23 ss. 2 and 10; SO 1914 c. 16 s. 4; Gregory Report, 203 and 192–8.

65 The Power Commission and Companies Transfer Act, SO 1924 c. 24 and Schedule A; The Central Ontario Power Act, SO 1930 c. 13; Gregory Report, 231.

4: The Ferguson Formula, 1925–43

1 Nelles, *Politics of Development*, 423; Denison, *People's Power*, 168 and 182.

2 OMEA, *Annual Meeting, 1931*, 14; OMEA, *Summer Convention, 1933*, 13.

3 *Canadian Who's Who, 1936–7*; Gaby, *Interesting Aspects of the Hydro System.*

4 Armstrong, *Politics of Federalism*, 72–3 and 161; Oliver, *Ferguson*, 176–7.

5 The HEPC had submitted a formal proposal to the IJC in 1921 to develop the Ontario portion of the St Lawrence. Although the United States was in favour of proceeding, King announced in May 1922 that the federal government would not permit the project to proceed at this time. According to Oliver, King was acting with caution knowing that the (yet to be nationalized) private power interests in Ontario were opposed to the control of the river's power developments by the HEPC and that interests in Quebec were opposed to both the power developments and canalization because these would increase Ontario's comparative advantage. Oliver, *Ferguson*, 174–6; Armstrong, *Politics of Federalism*, 162–5.

6 Oliver, *Ferguson*, 182.

7 Armstrong, *Politics of Federalism*, 166–7.

8 Shawinigan also sought to sell power to the HEPC but made its intention more attractive by offering a one-half interest in the project after forty years. No deal was reached. HEPC, *The Bulletin* 14, no. 2 (February 1927); Oliver, *Ferguson*, 184–6; Armstrong, *Politics of Federalism*, 167–8.

9 After Magrath had warned Meighen's minister of railways and canals, Henry Drayton, not to proceed without Ontario's consent, Drayton offered Ontario an

unqualified half-interest in the project and a one-third share of the annual water rental with the Quebec and federal governments receiving the other two-thirds. Armstrong, *Politics of Federalism*, 168–9; Oliver, *Ferguson*, 187.

10 Ontario, *Royal Commission re: the Ontario Power Service Corporation* (hereafter Latchford-Smith Report), 450, 452–3, and 457; Nelles, *Politics of Development*, 468–9.

11 Christie's preliminary report of 1 February 1927 contended, first, that the federal government's control of navigation did not give it the power to block power developments simply because canals were not then required and, secondly, that there was no constitutional justification for charging the cost of canals against power developments. His main report of 24 February analysed the specific mention of water power on canals in the third schedule of the Constitution Act, 1867 dealing with federal public works and property. He concluded that it was intended, in Oliver's words, to 'strike a balance sheet of the division of assets and debts as they existed in 1867' rather than 'to grant or define powers.' Oliver, *Ferguson*, 295–6; Bothwell, *Loring Christie*.

12 Oliver, *Ferguson*, 294–5; Armstrong, *Politics of Federalism*, 170.

13 Oliver, *Ferguson*, 352; Armstrong, *Politics of Federalism*, 171.

14 Neatby, *Mackenzie King, 1924–32*, 257–8; Regehr, *Beauharnois*, 46–9; Oliver, *Ferguson*, 353.

15 Regehr, *Beauharnois*, 26 and 46.

16 Armstrong, *Politics of Federalism*, 171–2; Oliver, *Ferguson*, 354–5; Regehr, *Beauharnois*, 65.

17 Regehr, *Beauharnois*, 80–1 and 84–5; OMEA, *Annual Meeting, 1932*, 4.

18 Nelles, *Politics of Development*, 423–4.

19 JCPC, *St Catharines* v. *Hydro-Electric Power Commission*.

20 Denison, *People's Power*, 199–200.

21 Saywell, *'Just Call Me Mitch,'* 73.

22 Gaby, *Interesting Aspects of the Hydro System*, 41–2.

23 The $125,000 figure allegedly represented 50 cents for every horsepower of the deal. Regehr, *Beauharnois*, 137–8.

24 OMEA, *Annual Meeting, 1932*, 9–11.

25 The royal commission, begun but not completed under Justice W.E. Middleton, was mandated only to inquire into the purchase of land on the Mississippi and Madawaska rivers by the HEPC (which involved a $50,000 payment to Aird) and the purchase of the Dominion Power and Transmission Company in 1930. *Canadian Annual Review, 1932*, 151.

26 The new accusation had not been made to the House of Commons committee in 1931 because Senator Haydon had been too ill to testify at that time. Ferguson, who had been appointed Canadian high commissioner in London, returned to

Canada immediately and denied the accusation before the Senate committee on 6 April. Ferguson's testimony was corroborated by Sweezey, although Sweezey's veracity was dubious given that he was negotiating with Bennett at this time to maintain control over Beauharnois. In the end, the Senate committee could only agree to reject Haydon's contention as unfounded. Haydon, it should be noted, had personally received $50,000 from Sweezey for ensuring that Beauharnois received the King cabinet's approval for the project on 8 March 1929. Regehr, *Beauharnois*, 156–7 and 44–5.

27 Justice Middleton resigned because of ill health on 16 April 1932, but not before recommending that the inquiry be extended to include the Beauharnois payment to Aird. Justice J.F. Orde took over the expanded mandate and concluded his investigation on 17 June but died before completing his report. Justices W.R. Riddell and G.H. Sedgewick took over on 19 August. Their report of 31 October found no impropriety in the HEPC's $50,000 payment to Aird, no relation between Beauharnois's payment to Aird and the HEPC's purchase of power, and that the purchase of the Dominion Power and Transmission Company was warranted and the price was appropriate. Ontario, *Royal Commission Appointed to Inquire into Certain Matters Concerning the Hydro-Electric Power Commission of Ontario*; *Canadian Annual Review, 1932*, 151–4; Latchford-Smith Report, 457–60; Saywell, '*Just Call Me Mitch*,' 98.

28 Gaby, *Trends of Electrical Demands*.

29 Ontario, *Journals, 1933*, 140–2; Latchford-Smith Report, 466–7 and 469–71; Abitibi Canyon Power Development Act, SO 1933 c. 1.

30 McKenty, *Hepburn*, 45; Canada, *Senate Debates, 1932–3*, 402–5; SO 1933 c. 1; Graham, *Meighen: Vol. III*, 53.

31 Roebuck, *Wreck of the Hydro*; HEPC, *Misleading Assertions ... Examined and Corrected*, 9; 'Hepburn Charges Hydro Assists Power Barons,' *Globe*, 24 August 1933, 1; HEPC, *Misleading Assertions ... Have Not Been Withdrawn*, 5–6.

32 Cooke, *Hydro Service ... Economic Aspects*.

33 OMEA, *Annual Report, 1934*, 4; HEPC, *The Bulletin* 21, no. 3 (March 1934), 83–7.

34 Ontario, *Journals, 1935*, 162; Graham, *Meighen: Vol. III*, 53–4.

35 HEPC, *Paid for Propaganda*, 15 and 27; Brady, 'Hydro-Electric Power Commission,' 339.

36 OMEA, *Summer Convention, 1934*; *Globe*, 7 July 1934.

37 *Globe*, 12 and 13 July 1934.

38 Ibid., 14 July 1934; Latchford-Smith Report, 463–4, 466, and 471; *Canadian Annual Review, 1934*, 188–9.

39 Latchford-Smith Report, 469–74; *Globe*, 17 July 1934; Graham, *Meighen: Vol. III*, 54–7; *Canadian Annual Review, 1934*, 188–9. Roger Graham gives a sympathetic but mistaken account of Meighen's transactions, believing the bonds were

exchanged at 60 per cent of the face value rather than 90 per cent. He does so without having consulted the royal commission report.

40 *Canadian Annual Review 1934*, 183–4; HEPC, *The Bulletin* 21, no. 9 (September 1934), 293; Nelles, *Politics of Development*, 478–9.

41 *Canadian Who's Who, 1936–7*; McDowall, *Brazilian Traction*, 279 and 333–4. Fleming confuses the position of controller with the statutory position of comptroller created in 1916 but never appointed, failing to note that the provision was removed from the Power Commission Act in 1927. See Fleming, *Ontario Hydro and Rural Electrification*, 167.

42 SO 1935 c. 54 s. 3; HEPC, *Annual Report, 1935*, xi; OMEA, *Annual Meeting, 1935*, 32 and 27–9.

43 OMEA, *Annual Meeting, 1935*, 20.

44 On forming the government, Hepburn approached Prime Minister Bennett about the scheme, only to discover that the Americans would not participate in negotiations on Niagara exclusive of the St Lawrence. Hepburn thereupon closed the diplomatic door on 'back to Niagara' when he responded in October 1934 that Ontario would not cooperate on the St Lawrence because the seaway's attendant power developments would be too expensive for Ontario to undertake during the depression. Armstrong, *Politics of Federalism*, 179–80.

45 Roebuck, 'Crisis in the Affairs of Hydro,' 164; Saywell, *'Just Call Me Mitch,'* 200. There was one example of a non-statutory municipal role in supply. The Hearst government had insisted on a popular sanction for the Queenston-Chippawa power development from the hydro municipalities in 1916 before it would pass the enabling legislation.

46 Saywell, *'Just Call Me Mitch,'* 200–1.

47 Ibid., 201–2.

48 Roebuck, 'Crisis in the Affairs of Hydro'; Gaby, *The Power Situation in Ontario.*

49 An Act to declare the law with respect to the Hydro-Electric Power Commission of Ontario and with respect to certain invalid Contracts, SO 1935 c. 53; Armstrong, *Politics of Federalism*, 181.

50 McCrimmon, *Report to the Chairman.*

51 Saywell, *'Just Call Me Mitch,'* 240–1.

52 Roebuck, 'Exhaustive Summary of Hydro Affairs'; HEPC, *Annual Report, 1935*, ix; *Canadian Annual Review, 1935–6*, 196–8.

53 Ontario, 'Select Committee ... Quebec Power,' 8–11 and 17; The Power Contracts Validation Act, SO 1936 c. 49; Saywell, *'Just Call Me Mitch,'* 246–7.

54 The chronology of the legal challenges was as follows: Beauharnois appealed for federal disallowance of the repudiation bill but was refused on 30 March 1936 on the grounds that it was a matter for the courts; Ottawa Valley filed a statement of claim in court, but Ontario's power to bar court challenges without the consent

of the attorney general was upheld on 3 June; the Ontario Court of Appeal agreed to rehear the case on 21 September; Beauharnois then began its legal action, but the judgment was reserved on 10 October pending the outcome of the Ottawa Valley appeal; Ottawa Valley won its case on 19 November; Beauharnois then won its case in the Supreme Court of Ontario on 13 January 1937; the government responded by appealing the Ottawa Valley decision to the Judicial Committee and the Beauharnois decision to the Ontario Court of Appeal. *Canadian Annual Review, 1935–6*, 199–200; *Canadian Annual Review, 1937–8*, 152–3.

55 The Power Commission Declaratory Act, SO 1937 c. 58; The Power Commission Amendment Act, SO 1937 c. 59; The Privy Council Appeals Amendment Act, SO 1937 c. 62.

56 Saywell, *'Just Call Me Mitch,'* 286–7.

57 Ontario, 'Select Committee ... Quebec Power,' 12–13.

58 Ibid.; SO 1937 c. 58; SO 1937 c. 59; The Power Contracts Validation Act, SO 1937 c. 61.

59 OMEA, *Annual Meeting, 1937*, 4.

60 *Canadian Annual Review, 1937–8*, 154; Saywell, *'Just Call Me Mitch,'* 302 and 336; Ontario, 'Select Committee ... Quebec Power,' 15 and 27.

61 Ontario, 'Select Committee ... Quebec Power,' 16–17.

62 OMEA, *Annual Meeting, 1938*, 13, 3, and 5; HEPC, *The Bulletin* 25, no. 10 (October 1937), 301–5. Saywell mistakenly writes that Arthur St Clair Gordon, MPP, was made the third commissioner. Saywell, *'Just Call Me Mitch,'* 361.

63 The Power Contracts Validation Act, SO 1938 c. 27; Ontario, 'Select Committee ... Quebec Power,' 17–18; Armstrong, *Politics of Federalism*, 186–8.

64 Hogg, 'Quebec Power Contracts Settlement,' 52, 57, and 61.

65 OMEA, *Annual Meeting, 1938*, 3, 5, 15, and 24; Houck, 'Municipalities and the Hydro Scheme,' 37–40; OMEA, *Annual Meeting, 1939*, 4 and 26.

66 Ontario, 'Select Committee .. Quebec Power,' 4, 17, 33–4, 37, and 40.

67 Hogg, 'Mutual Interdependence'; The Power Control Act, SO 1939 c. 8 (second session).

68 HEPC, *Annual Report, 1941*, viii–ix; HEPC, *Annual Report, 1942*, xi; The Ottawa River Water Powers Act, SO 1943 c. 21; Saywell, *'Just Call Me Mitch,'* 435 and 483–4; Armstrong, *Politics of Federalism*, 194.

69 *Toronto Daily Star*, 19 February 1943.

5: One-Party Dominance, 1943–63

1 The HEPC's affairs were the thirteenth point of the platform. This point also included equitable power rates, accelerated rural electrification and the removal of a special rural power charge. *Globe and Mail*, 9 July 1943.

2 OMEA, *Annual Meeting, 1944*, 24, 28–34, 40, and 44; SO 1944 c. 51 s. 1. Robert Macaulay, who later became a HEPC commissioner and the minister of energy resources, postulates that the council was one of a number of ideas with which the Conservatives experimented after being so long out of power but was found to be either impractical or unnecessary. Macaulay, interview. References to people interviewed and corresponded with will only be made where their name does not appear in the text. A complete list of those interviewed and the dates of the interviews is included with the bibliography.

3 Strike had been on the Bowmanville commission since 1932 and had been mayor of Bowmanville, 1932–5.

4 OMEA, *Annual Meeting, 1946*, 6 and 23.

5 Hogg, 'Hydro's Postwar Challenges'; SO 1946 c. 73 s. 1.

6 *Globe and Mail*, 28 January 1947, 23.

7 Ibid., 1 January 1947, 1, and 28 January 1947, 28; *Toronto Daily Star*, 25 January 1947.

8 *Evening Telegram*, 24 January 1947, 1; *Globe and Mail*, 24 January 1947, 4, and 28 January 1947, 23; *Toronto Daily Star*, 24 January 1947, 1, and 25 January 1947, 1–2.

9 The radio address was reprinted in the *Globe and Mail*, 28 January 1947, 23, and the *Toronto Daily Star*, 28 January 1947, 10.

10 *Evening Telegram*, 28 January 1947; *Globe and Mail*, 29 January 1947.

11 *Toronto Daily Star*, 28, 29, and 30 January 1947.

12 *Globe and Mail*, 3 February 1947, 3.

13 The agreement was reported on 31 January 1947, but its details were not made public until 10 February 1947, after the Ontario cabinet had given its approval. *Globe and Mail*, 31 January 1947, 1, and 11 February 1947, 1.

14 OMEA, *Annual Report, 1947*, 9–10, 23–9, 16–17, and 50.

15 Ibid., 6–8; HEPC, *Interim Report on a Proposal to Standardize Frequency*, 13.

16 OMEA, *Annual Report, 1947*, 25, 32–43, and 46.

17 Gordon, *Report upon the Proposed Organization Plan*, 2.

18 Ibid. (emphasis added). The report recommended an internal reorganization of the HEPC to meet its particular functional and regional needs and requirements. These matters were dealt with in greater detail in subsequent reports. Challies commissioned the report on 5 February 1947. This was three weeks prior to Hogg's official resignation on 28 February and Challies's own official assumption of the acting-chairmanship.

19 SO 1947 c. 78.

20 Denison, *People's Power*, 236. The provision of the Power Commission Act which permitted the commissioners to appoint the chief engineer and other officers was amended in 1949 to add the function of general manager. SO 1949 c. 73 s. 1.

21 OMEA, *Annual Meeting, 1947*, 24 and 34; HEPC, *Summary of the Reports on Frequency Standardization*, 15.
22 OMEA, *Annual Meeting, 1947*, 48–51.
23 Ontario, *Journals, 1948*, 8, 53, and 89–90; SO 1948 c. 69 s. 2; OMEA, *Minutes of the Executive Committee*, 5 April 1948.
24 OMEA, *Special General Meeting, 1948*, 3; Legislature of Ontario, *Debates*, 16 April 1948; Graham, *Frost*, 133.
25 OMEA, *Special General Meeting, 1948*, 6, 11–15, and 20; OMEA, *Minutes of the Executive Committee*, 3 May and 21 June 1948. The OMEA used a complicated weighted voting scheme. After hours of tabulation, the larger municipal hydros, who had the most votes, ensured that the resolution passed. Hutcheson, *Memoirs*, 218–19.
26 SO 1949 c. 73. ss. 2 and 7; SO 1950 c. 55 ss. 1 and 2; RSO 1937 c. 62 s. 7.
27 Graham, *Frost*, 212–13 and 15.
28 HEPC, *Hydro News* 38, no. 3 (March 1951), 15.
29 The Niagara Development Act, SO 1951 c. 55; The Niagara Development Agreement Act, SO 1951 c. 56.
30 SO 1951 c. 67 s. 1; Graham, *Frost*, 214–15; HEPC, *Hydro News* 38, nos. 7 and 8 (July-August 1951), 18–19.
31 Graham, *Frost*, 213–15; HEPC, *Annual Report, 1951*, vii. Strike used these words in private conversation with his son.
32 HEPC, *Hydro News* 38, nos. 7 and 8 (July-August 1951), 18–19.
33 Ibid., 39, no. 3 (March 1952), 9. There are no minutes for either the OMEA executive or annual meetings for this period.
34 Denison, *People's Power*, 246.
35 OMEA, *Annual Meeting, 1953*, 9; HEPC, *Hydro News* 39, no. 3 (March 1952), 9.
36 The International Rapids Power Development Agreement Act, SO 1952 c. 42; The St Lawrence Development Act (No. 2), SO 1952 c. 3 (second session). The last act repealed SO 1952 c. 100, which had been passed on 10 April in the first session.
37 AECL, *Annual Report*, 1952–3; Bothwell, *Nucleus*, 144–5 and 189–91; Eggleston, *Canada's Nuclear Story*, 308–10; Hodgetts, *Administering the Atom for Peace*, 107–8. The Power Commission Act was amended in 1955 to include atomic power under its definition of power. SO 1955 c. 62 s. 1.
38 Graham, *Frost*, 214.
39 OMEA, *Annual Meeting, 1955*, 39–40. A.W. Manby and Otto Holden, both long-time HEPC employees, split Hearn's former duties, becoming general manager and chief engineer, respectively.
40 Ibid., 42; Legislature of Ontario, *Debates*, 15 March 1955.
41 SO 1955 c. 62 ss. 2, 3, 4, and 5.

42 Bothwell, *Nucleus*, 238; Cook, *Massey on the Brink*, 17; Graham, *Frost*, 373.
43 OMEA, *Annual Meeting, 1957*, 81–2.
44 Ibid., 77; OMEA, *Annual Meeting, 1958*, 9, 13, 18–19, 50, and 51–6; HEPC, *Annual Report, 1958*, ix.
45 Graham, *Frost*, 374.
46 Although Macaulay is not listed as having presented a brief to the committee, he did so, he recalls, at a public hearing at the invitation of the committee. Ontario, *Report of the Committee on the Organization of Government in Ontario* (hereafter Gordon Report), 89 and 87; Duncan, *Brief to the Committee on the Organization of Government in Ontario*, 6–7.
47 Ontario, *Journals, 1959*, 4.
48 Legislature of Ontario, *Debates*, 10 February 1959.
49 Graham, *Frost*, 374–5.
50 OMEA, *Annual Meeting, 1959*, 49, 54, 66, 70, and 96.
51 The Department of Energy Resources Act, SO 1959 c. 26.
52 The Crown Agencies Act, SO 1959 c. 22; Ontario, *Bills, 1959*.
53 MacDonald, *The Happy Warrior*, 302–3. Macaulay authored a book on the economic development potential of nuclear power, *A Layman's Guide to Nuclear Power*, in 1959. He would write a second, *The World of Energy*, in 1961.
54 Gordon Report, 80–1.
55 Ibid., 38.
56 The Energy Act, SO 1960 c. 30; The Ontario Energy Board Act, SO 1960 c. 73; The Ontario Fuel Board Amendment Act, SO 1958 c. 71 s. 1.
57 Ontario, *Select Committee on Administrative and Executive Problems*, 5–6 and 15. A second interim report was delivered on 19 March 1962, but no recommendations were made regarding the HEPC. There was no final report before the committee was dissolved with the 1963 election.
58 Ontario, *Bills, 1960–1*; Legislature of Ontario, *Debates*, 28 March 1961, 2217–49; OMEA, *Annual Meeting, 1961*, 8–9 and 59–63.
59 OMEA, *Annual Meeting, 1961*, 97–9; Don Cliff, interview; Ontario, *Journals, 1960–1*, 74 and 149.
60 OMEA, *Annual Meeting, 1961*, 75–7.
61 Ontario, *Journals, 1961–2*, 29.
62 Manthorpe, *Power and the Tories*, 91.
63 OMEA, *Annual Meeting, 1962*, 74–5 and 79.
64 The Power Commission's Systems Consolidation Act, SO 1961–2 c. 107.
65 Ontario, *Journals, 1961–2*, 29, 117, 120, and 123; SO 1961–2 c. 106 s. 2; OMEA, *Annual Meeting, 1962*, 87–8; OMEA, *Annual Meeting, 1963*, 11–12. According to Don Cliff, Merson was a complicated individual, one who did not like change and likely felt Macaulay had an ulterior motive.

66 OMEA, *The OMEA – Ontario Hydro, the People's Power – and Government*; OMEA, *Annual Meeting, 1963*, 16–17.
67 OMEA, *Annual Meeting, 1963*, 16–17, 104–5, and 108.
68 HEPC, *Annual Report, 1963*, vii; Leonoff, correspondence.

6: The Age of Government Modernization

1 Ontario, *Journals, 1964*, 6–7; Legislature of Ontario, *Debates*, 30 January 1964, 312–14; OMEA, *Annual Meeting, 1964*, 65–6; OMEA, *Annual Meeting, 1965*, 13; HEPC, *Annual Report, 1964*, iv.
2 Ontario, *Third Report of the Select Committee on the Municipal and Related Acts*, 122 and Appendices E, F, and G; OMEA, *Annual Meeting, 1965*, 6–7; Ontario, *Fourth and Final Report of the Select Committee on the Municipal and Related Acts*, 153–4, 169, 173, 176, and 185; Jones, *Final Report and Recommendations*, 36, 45, 47, and 61; OMEA, *Annual Meeting, 1966*, 8–12, 22, 59–60, and 62.
3 OMEA, *Annual Meeting, 1966*, 8; HEPC, *Hydro News* 52, no. 12 (December 1965), 16–17.
4 HEPC, *Hydro News* 53, no. 4 (April 1966), 10; Leonoff, correspondence.
5 The formula was the result of a two-year review which included consultation with local hydros and the OMEA. In answer to Frame's complaint, McMechan responded that the Power Costing Committee had been aware of 'all aspects of the situation' and that 90 per cent of the hydro municipalities had decided in favour of the new system. HEPC, *Annual Report, 1965*, iv; OMEA, *Annual Meeting, 1966*, 70–5, 33–5, and 62.
6 SCC, *Nepean* v. *Ontario Hydro*, 192; TFH, *Report One*, 15–17.
7 OMEA, *Annual Meeting, 1967*, 5–6, 38, 42, 32–3, and 14.
8 Ontario Committee on Taxation, 'Provincial Government Enterprises,' chapter 36 (hereafter Smith Report). On the first recommendation, the committee believed that the guarantee distorted the credit ratings of both the province and the HEPC and that the commission could 'reasonably be expected to meet the competitive test of the capital market' (paragraphs 22–3, 420). On the second, the committee recognized that depreciation and other allowances would, 'in all probability,' not produce a taxable income but felt the tax calculation had merit because it revealed the degree to which the HEPC was 'liable' for taxation (paragraphs 20–1, 419–20). On the third, the committee objected to the sinking fund because it was not in accord with 'generally accepted' accounting practices (paragraphs 13–18 and 418–19).
9 The most controversial of these recommendations related to control over surpluses. The committee felt that where a high level of revenue retention exists, which was especially the case for hydro commissions, present users were in fact

subsidizing future users. The committee reasoned that since the municipal councils must cover any losses incurred, they should be able to determine the level and administration of the surplus. Smith Report, 'Reconciling Structure with Finance,' chapter 23, paragraphs 151, 54, 544, and 514–15; 'Local Non-Tax Revenue,' chapter 17, paragraphs 43–54 and 341–7.

10 On those recommendations affecting the HEPC, Gathercole felt that it was a fallacy to subject a public, at-cost enterprise to the same taxation as a private, for-profit corporation; that the HEPC met its full tax responsibilities; and that there was no useful purpose in adopting private sector accounting practices. On those affecting the municipal hydros, he stated that supervision should remain with the HEPC. OMEA, *Annual Meeting 1968*, 79–80. The Smith Committee's recommendations would be tempered by those of a select committee of the legislature which reported 16 September 1968. On those affecting the HEPC, it rejected the need for private sector accounting practices, felt payment of corporate income tax needed more study, and did not endorse the removal of the government guarantee of HEPC debt. On those affecting local hydro commissions, it concluded that they should be subject to only 'municipal' taxation but concurred that municipal councils should control their surpluses and that the Department of Municipal Affairs should supervise their finances. On regional government, the White Committee also did not list hydro as an upper-tier function. Ontario, *Select Committee Report on the Ontario Committee on Taxation*, 293–5, 101–3, and 145–50.

11 The transitory factors Gathercole noted were 'high interest rates, wages and salaries; rising prices for materials and equipment; and the transition to predominantly thermal power generation and a far more sophisticated and expensive transmission system.' OMEA, *Annual Meeting, 1968*, 25 and 76–8; HEPC, *Annual Report, 1967*, iv.

12 Ontario, *Journals, 1968*, 9; OMEA, *Annual Meeting, 1968*, 91–2 and 54–6; OMEA, *Annual Meeting, 1969*, 82; Ontario, *Design for Development [Phase One]*. On announcing the second phase of Design for Development in 1968, Premier Robarts, for his part, stated that the economic initiatives of phase one and the local government reviews under way had come to be interrelated, declaring that the moment had arrived for both to be pursued under regional government. McKeough added that regional government would not be instituted uniformly across the province as the OMEA had feared. Ontario, *Design for Development: Phase Two*.

13 The HEPC concurred with the flexibility of this policy position, according to Gordon, because the impetus for services controlled by regional councils, planning on a larger scale, was not a problem because the HEPC already performed this function. OMEA, *Annual Meeting, 1969*, 1, 4, 87, and 119–20; McDougall, *Robarts*, 229–30; OMEA, *Annual Meeting, 1970*, 94, 96, and 133.

14 While Gordon recalls Gathercole's pronouncements as appropriate for the time, he remarks that where economic spin-offs might have been an important consideration for the government, the HEPC was primarily concerned with establishing a sound supply industry. OMEA, *Annual Meeting, 1969*, 71; HEPC, *Annual Report, 1968*, iii–v.
15 OMEA, *Annual Meeting, 1969*, 44–6; Frame, interview.
16 OMEA, *Annual Meeting, 1970*, 2; HEPC, *Hydro News* 56, no. 7 (June/July 1969), 22; Leonoff, correspondence; McDougall, *Robarts*, 233.
17 COGP, *Interim Report One*, Appendix 1, 37 and 8–9.
18 HEPC, *Annual Report, 1969*, iv–v; OMEA, *Annual Meeting, 1970*. On the lack of responsiveness of municipal hydro commissions to the public, see, for example, OMEA, *Annual Meeting, 1969*, 77–80.
19 HEPC, *Annual Report, 1970*, iii–iv.
20 OMEA, *Annual Meeting, 1971*, 42–3, 94–5, and 89–94.
21 Ibid., 76–83.
22 COGP, *Interim Report Two*; COGP, *Interim Report One*, 8–9; Szablowski, 'Policy-Making in the Cabinet,' 117–18; Silcox, 'ABC's of Ontario,' 138–9.
23 See COGP, *Interim Report Four*, 13–17.
24 Legislature of Ontario, *Debates*, 22 July 1971; Advisory Committee on Energy, *Energy in Ontario; Volume One*; OMEA, *Annual Meeting, 1972*, 21. The Advisory Committee was not officially appointed until 18 August 1971.
25 The Department of the Environment Act, SO 1971 c. 63; The Environmental Protection Act, SO 1971 c. 86; COGP, *Interim Report Three*.
26 The Committee met and concluded that it would be irresponsible to set rates below 7 per cent before Frame announced, to the surprise of those assembled, that the government did not want an increase over 5 per cent. OMEA, *Power Costing in Ontario*; OMEA, *Annual Meeting, 1972*, 32–3.
27 OMEA, *Annual Meeting, 1972*, 21 and 32–3.
28 COGP, *Interim Report Three*, 1–3 and 14–26, and rec. 8.1–8.6.
29 HEPC, *Annual Report, 1971*, 2–4.
30 Ibid.; HEPC, *Statistical Yearbook, 1971*; SO 1918 c. 14 s. 3; SO 1950 c. 55 s. 2.
31 OMEA, *Annual Meeting, 1972*, 91–7.
32 Ibid, 43–5.
33 Ibid., 115–18, 143, and 146–9.
34 HEPC, *Hydro News* 59, no. 4 (April 1972), 16; OMEA, *Annual Meeting, 1973*, 4.
35 The Government Reorganization Act, SO 1972 c. 1 ss. 3 and 73; SO 1916 c. 19 s. 4; SO 1949 c. 73 s. 2. As Bill 27, the statute had been debated at length in Committee of the Whole on 30 March 1972, but its effect on the HEPC was not raised as an issue. Legislature of Ontario, *Debates*, 30 March 1972, 740–68. There was no provision for the annual report until 1916; it was submitted

directly to the cabinet until 1949, then to the provincial secretary (as distinct from the new policy ministers) until 1972. From the HEPC perspective, Gordon and Taylor, who would become president and chairman of Ontario Hydro respectively, remark that the change only meant that the HEPC was responsible 'through the minister' rather than 'to the minister.' Doern writes that this 'statutory ambivalence is of considerable import and often politically convenient for both parties.' Doern, *Canadian Nuclear Industry*, 165.

36 Ontario, *Design for Development: Phase Three*.

37 Letter from Blake, Cassels & Graydon to TFH, 13 June 1972, and letter from Blake, Cassels & Graydon to TFH, 19 September 1972. The share certificate scheme was recommended to the task force between the first and second letters, according to Muncaster, but was presented in the second letter. The substance of this legal opinion appears in Task Force Hydro (TFH), *Report One*, 60–3.

38 The Municipal Electric Association, successor to the OMEA, refuses to release the legal opinion, still considering its content to be confidential. Jennings, correspondence.

39 Solandt Commission, *Interim Report*; Solandt Commission, *Closing the Gap*.

40 TFH, *Report One*, rec. 1.18, 47–9, and Appendix V, 80–1. Reynolds does not remember the Crown Agency Act ever coming to the attention of TFH, the COGP, or any law reform commission. Macaulay states the exemption would have been nullified if the government passed a statute calling the HEPC a crown corporation.

41 In addition, TFH had also been motivated by the need to fit the HEPC into the role of a crown corporation as this would be defined by the COGP, recognizing that it would not fit the role of a commission. TFH, *Report One*, rec. 1.18, 47–9; Morrison, *Government and Parliament*, 282–3. The COGP would not make its agency recommendations until March 1973. See COGP, *Report Nine*.

42 TFH, *Report One*, rec. 1.20, 49–52, and Appendix V, 80–1.

43 Ibid., rec. 1.1, 35–6 and Appendix V, 80–1.

44 Ibid,. rec. 1.21 and 1.22, 53–4.

45 Ibid., rec. 1.14–1.17, 44–5.

46 Ibid., rec. 1.5–1.9, 40–2.

47 Ibid., rec. 1.31, 60–3.

48 Ibid., rec. 1.25, 1.26, and 1.29, 56–9, and Appendix V, 80–1.

49 OMEA, *Submission to the Premier*.

50 Ontario, *Statement by Premier Davis re: First Report of TFH*.

51 Solandt, *Interim Report*, 31–3; TFH, *Report One*, rec. 1.10 and 1.12, 43; TFH, *Report Two*, rec. 2.4, 2.5 and 2.6, 61–6.

52 Advisory Committee on Energy (ACE), *Energy in Ontario: Volume One*, rec. 00:1,

i–ii and Appendix F, 37–9. A second volume was later published by the committee as the main report, with volume one serving as the committee's summary of findings and vehicle for its principal recommendations. ACE, *Energy in Ontario: Volume Two*.

53 ACE, *Volume One*, rec. 00:1, i–ii.

54 In presenting this recommendation, the committee did recognize that the government might be reluctant to create a new ministry so soon after the COGP round of reorganization that established the Provincial Secretary for Resources Development portfolio. As a fall-back position, it recommended the energy commission be made permanent. ACE, *Volume One*, rec. 00:3–00:7, iii–vi.

55 ACE, *Volume One*, Appendix E, 36.

56 TFH, *Response to OMEA Statements*; Frame, interview.

57 TFH, *Response to OMEA*; McKeough, *Structure of Energy Administration*; Manthorpe, *Power and the Tories*, 250–3.

58 HEPC, *Hydro News* 60, no. 2 (March–April, 1973), 13.

59 TFH, *Report Three*, rec. 3.2–3.18, 2, 43–59. According to Reynolds, the growth assumptions underlying the nuclear study, which were offered without explanation, were premised on the historical rate of growth in demand, rather than looming energy shortages. Dillon believes the opposite because society had been brainwashed by the oil and gas industry. Frame suggests both were contributing factors.

60 OMEA, *Annual Meeting, 1973*, 4–7.

61 In his introductory remarks, Davis cautioned that commission recommendations 'do not automatically become government policy' and that he had deferred judgment on TFH's recommendations to allow for discussion. Ibid., 81–7.

62 Ibid., 83, 88–90; *Design for Development: Phase Three*.

63 OMEA, *Annual Meeting, 1973*, 78 and 115.

64 In response to specific TFH recommendations, Gathercole stated that there must be a convincing reason for hydro to remain a lower-tier responsibility and that the relationship between the HEPC and the municipalities was 'vital and must be maintained.' OMEA, *Annual Meeting, 1973*, 91–4.

65 Before taking questions from the floor, Muncaster clarified four points of misunderstanding. He explained that in recommending that the HEPC be responsible for the 'total delivery system,' TFH was giving recognition to the fact that the Power Commission Act already gave it the power to regulate municipal rates and other matters; he stated that TFH had mistakenly been seen to be calling for profits to be generated when in fact it favoured generating more funds internally just as the OMEA had with a capital levy; on regional government, he explained that upper-tier hydro was preferable because of greater economies of scale and more uniform rates; and finally, he stated that TFH would recommend rate

review because the present consultation between the HEPC and OMEA did not provide for public participation. Ibid., 100–5 and 68–7.

66 Ibid., 107–15.

67 The OMEA also called for a supplementary paper clarifying TFH's recommendations, an idea which Davis rejected on 16 April 1973. OMEA, *Annual Meeting, 1974*, 3.

68 The COGP also recommended that all appointments to agencies be made by cabinet, and only for limited terms, and that all directors of corporations be given freedom to focus on economic performance, neither of which reinforced the OMEA'scall for half the directors. Of related interest, the COGP recommended that all agencies report to the legislature through a minister. COGP, *Report Nine*, rec. 12.16, 12.17, 12.21, 12.22, 12.24, and 12.2, 47–50, and 38.

69 Ontario, *Journals, 1973*, 5; McKeough, *Structure of Energy Administration*. In the knowledge that politicians wanted the high-profile relationship recommended by the advisory committee, Dillon informed McKeough of the importance of the delicate balance TFH had sought between corporate autonomy and policy direction for both the HEPC and the government. Macaulay, who also favoured the TFH model over that of the advisory committee, remembers his advice being on the order of finding a politically acceptable solution.

70 McKeough, *Structure of Energy Administration*, 10–14.

71 TFH, *Report Four*, rec. 4.9, 33. On the issue of strategically releasing reports to maximize their chance of implementation, James Fleck, executive director of the COGP, remarks that if TFH's reports were not released strategically like the COGP's, 'they should have been.' On the matter of releasing COGP's reports strategically, see COGP, *Report Nine*, 46–7.

72 Frame also dissented on a subsidiary recommendation for the HEPC to retain all surpluses for use at its own discretion, one effect of which would be a discontinuance of the practice of paying returns on the sinking fund payments. Muncaster feels the municipal hydro commissions were hypocritical in wanting the returns while not wanting to make up for short-falls in other years. TFH, *Report Four*, rec. 4.1–4.7, 4.14 and 4.16, iii, 16–31, 33, 39–41, and 55.

73 Ibid., rec. 4.17, iii and 56–8.

74 Ibid., rec. 4.27, 100–2.

75 Ibid., rec. 4.18, 72–3.

76 *Globe and Mail*, 28 April 1973, 1 and 4; Manthorpe, *Power and the Tories*, 266–73.

77 Gathercole maintains that the HEPC's chief architect, Ken Candy, had led him to believe that the 'other builders had been offered the opportunity to tender.' Davis accepts this view, which was forwarded in a subsequent select committee inquiry. *Globe and Mail*, 30 April 1973, 1 and 5; Manthorpe, *Power and the Tories*, 266–73; Ontario, *Report of the Select Committee on the Hydro Head Office Building, 1973*.

78 Legislature of Ontario, *Debates*, 30 April 1973, 1381; Ontario, *Journals, 1973*, 63–4.

79 HEPC, *Annual Report, 1972*, 4–7.

80 Legislature of Ontario, *Debates*, 7 June 1973, 2779–85; HEPC, *Hydro News* 60, no. 4 (July/August 1973), 22–5; McKeough, *Structure of Energy Administration.*

81 The Power Commission Amendment Act, SO 1973 c. 57; The Ministry of Energy Act, SO 1973 c. 56; The Ontario Energy Board Amendment Act, SO 1973 c. 55. To enact these statutes, two others were repealed or amended for conformity: The Power Control Repeal Act, SO 1973 c. 58, and The Power Commission Insurance Amendment Act, SO 1973 c. 59. TFH would deliver a fifth report after the legislation had passed.

82 SO 1973 c. 57 ss. 1–5. The indemnification heeded the OMEA's concern that removal of the leave requirement, as recommended by TFH, would introduce the 'adversary process' into Ontario Hydro's affairs. TFH, *Report One*, rec. 1.11, 43; OMEA, *Submission to the Premier.*

83 SO 1973 c. 57, s. 7.

84 Ibid., ss. 3 and 4.

85 The Crown Agency Act, RSO 1980 c. 106 s. 3.

86 SO 1973 c. 56 ss. 4 and 8.

87 SO 1973 c. 55 ss. 2 and 12; The Energy Amendment Act, SO 1970 c. 61.

88 SO 1973 c. 55 s. 12.

7: An Epilogue for Ontario Hydro

1 The royal commission was formally appointed on 17 July 1975. Grossman, 'Statement on the Long-Range Planning of Ontario's Electric Power System'; Ontario, *Royal Commission on Electric Power Planning: Volume 1*, Appendix A, 187–8 (hereafter Porter Commission); *Canadian Annual Review, 1975*, 130; *Canadian Annual Review, 1974*, 171; Ontario Hydro, *Annual Report, 1975*, 2.

2 Ontario, *Commission on the Legislature*; *Canadian Annual Review, 1975*, 130–1.

3 *Canadian Annual Review, 1976*, 183; Ontario, *Final Report of the Select Committee Investigating Ontario Hydro.*

4 Porter Commission, 7; *Canadian Annual Review, 1977*, 135–6; Environmental Assessment Act, SO 1975 c. 69. For the select committee's terms of reference, see Ontario, *Select Committee on Ontario Hydro: Report on Uranium Contracts*, Appendix A.

5 *Canadian Annual Review, 1978*, 122.

6 Porter Commission, 27, 29, and 37, rec. 3.1 and 3.6; Ontario, *Response to the Royal Commission on Electric Power Planning.*

7 These promises fell under the auspices of a new Board of Industrial Leadership and Development. Ontario Hydro, *Annual Report, 1981*, 4.

8 'Reality Includes Nastiness,' *Globe and Mail*, 24 April 1981; 'Hydro Probe Victim of Majority,' ibid., 12 May 1981, 4.

9 SO 1981 c. 16 s. 3; *Canadian Annual Review, 1982*, 155–6 and 214.

10 SCC, *Nepean* v. *Ontario Hydro*, 222, 243, and 198.

11 Speirs, *Out of the Blue*, 82, 91, and 125.

12 *Canadian Annual Review, 1986*, 250–1.

13 Ontario, *Journals, 1987–9*, 13; 'Hydro Harnessed after the Horsepower Has Bolted,' *Toronto Star* 31 October 1987, D1; 'Hydro Chairman Will Resign Next Month,' ibid., 2 December 1987, A29.

14 SO 1989 c. 36; Ontario, *Budget, 1989*, 27 and 29.

15 SO 1989 c. 53 ss. 24 and 26.

16 Ibid., ss. 8 and 2.

17 'Peterson Denies Advance Approval of Hydro Plan,' *Globe and Mail*, 21 December, 1989; Ontario Hydro, *Providing the Balance of Power*; *Globe and Mail*, 26 January 1993; Gagnon and Rath, *Not Without Cause*, 94 and 103.

18 Ontario, *Journals, 1990–1*, 15; 'Future Shock,' *Toronto Star*, 23 August 1992; 'Rae Makes Pick for Top Hydro Post,' *Globe and Mail*, 25 April 1991; 'New Hydro Chief Controversial,' ibid., 17 May 1991; 'Ontario Hydro, Critics in Open Environmental Battle,' ibid., 23 April 1991; ibid., 30 May 1991; 'Is Hydro Working?,' *Toronto Star*, 13 October 1991.

19 SO 1992 c. 10 ss. 1, 8, and 2.

20 'Hydro Defers Nuclear Plants,' *Globe and Mail*, 16 January 1992; '44% Jump in Hydro Rates,' ibid., 17 July 1991.

21 'Ontario Energy Minister Resigns in Wake of "Serious Allegation,"' *Globe and Mail*, 14 February 1992; 'NDP Likely to Veto Nuclear Plants,' ibid., 10 September 1990; 'Chairman's Resignation Shocks Ontario Hydro,' ibid., 1 August 1992; 'Hydro Official Got $1.2 Million,' ibid., 3 November 1992.

22 'Strong Calls Ontario Hydro "a Corporation in Crisis,"' ibid., 10 December 1992; '12,000 at Hydro Offered Severance,' ibid., 14 December 1992; 'Ontario Hydro to Cancel Big Deal,' ibid., 17 December 1992; 'Small Power Developers Left in Lurch,' ibid., 24 December 1992; 'Ontario Hydro to Halt Review on Expansion,' ibid., 26 January 1993.

23 Ontario Hydro, *Hydro 21*.

24 Strong, 'Hydro in Ontario: Strengthening the Partnership'; and Ontario Hydro, *Hydro in Transition*.

25 Ontario Hydro, *The New Ontario Hydro*.

26 Ibid.; 'Strong's Jungle Plan Awkward for the NDP,' *Globe and Mail*, 19 May 1994; 'Hydro Expands Overseas,' ibid., 13 July 1994; 'Ontario Hydro's Offshore

Investments to be Reviewed,' ibid., 20 September 1994; OEB, *Ontario Hydro International*, rec. 3.1 and 3.3.

27 The MEA examination was initiated during the Environmental Assessment Board's hearings on Hydro expansion because evidence was being heard on privatization and decentralization. MEA, *Phase I Report*.

28 'Ontario Hydro Facing Huge Loss,' *Globe and Mail*, 13 July 1993; MEA, *Phase II Report*; 'Ending Ontario Hydro's Monopoly (I), (II) and (III),' *Globe and Mail*, 11, 12, and 13 November 1993; 'Hydro Subject of Informal Privatization Talks,' ibid., 20 July 1993; 'Ontario Hydro not for Sale,' ibid., 21 Aug 1993.

29 Strong, 'Hydro in Ontario: Forging a New Partnership.'

30 Ontario Hydro, *Annual Report, 1993*; Ontario Hydro, *Annual Report, 1992*; Ariss, correspondence and interview; 'Ontario Hydro Gushes Red Ink,' *Globe and Mail*, 18 January 1994.

31 MEA, *Phase III: Interim Report*; 'Ontario Hydro: Hostile Takeover Target,' *Globe and Mail*, 1 March 1994; MEA, *Final Phase III Report*.

32 'Ontario Hydro Restricts Access to System,' *Globe and Mail*, 17 June 1994; Ontario Hydro, *Challenges and Choices*, 7–8, 25–6, 17–18, and 30; 'Hydro Stock Sale Could Net $5 Billion,' *Globe and Mail*, 28 June 1994.

33 'Ontario Hydro Seeks More Freedom,' *Globe and Mail*, 21 July 1994.

34 'Strong Cuts Pay in Big Way,' ibid., 13 September 1994; 'Ontario Hydro Gives Rate Relief,' ibid., 18 October 1994; SO 1994 c. 31 s. 1; 'New Hydro Head Braces for Competition,' *Globe and Mail*, 18 January 1995.

35 Ontario Hydro, *Annual Report, 1994*; Progressive Conservative Party of Ontario, *Common Sense Revolution*, 14–15; 'Tories Eye Privatization Plans,' *Globe and Mail*, 2 May 1995.

8: Conclusion

1 Tuohy, *Policy and Politics*, xvii.

2 Heard, *Canadian Constitutional Conventions*.

3 Tuohy, *Policy and Politics*, 347.

4 Ibid., 347–8.

Bibliography

Interviews and Correspondence

(Except where telephone interviews are noted, all the interviews were taped and transcribed.)

Ariss, D.G. Manager of accounting policy, Ontario Hydro. Correspondence, 8 November 1994, and interview, 21 December 1994 (by telephone).

Cliff, Don H. Son of D.P. Bud Cliff (HEPC commissioner 1956–69, president of the Ontario Municipal Electric Association 1950–1, secretary-treasurer of the OMEA 1953–66) and formerly of consumer service department, Ontario Hydro. Interview, Burlington, Ontario, 9 October 1991.

Cronyn, John B. Chairman of the Committee on Government Productivity 1966–73, and senior vice-president and director of John Labatt. Interview, London, Ontario, 2 March 1992.

Davis, Hon. William G. HEPC commissioner 1961–2 and premier of Ontario 1971–85. Interview, Toronto, Ontario, 17 August 1992.

Dillon, Richard M. Executive director and member of Task Force Hydro 1971–3, member of the Advisory Committee on Energy 1971–3, deputy minister of energy 1973–6, and deputy provincial secretary for resources development 1976–9. Interviews, Toronto, Ontario, 5 March 1992 and 7 July 1992.

Fleck, James D. Executive director of the Committee on Government Productivity 1970–1, chief executive officer, office of the premier 1972–3, and secretary to the cabinet 1974–5. Interview, Toronto, Ontario, 24 February 1992.

Frame, Andrew. President of the Ontario Municipal Electric Association 1970–1, member of Task Force Hydro 1971–3, director of Ontario Hydro 1974–5, and senior advisor, utilities and operations, Ministry of Energy, 1975–92. Interviews, Toronto, 14 February and 12 March 1992, and Burlington, Ontario, 21 July 1992.

Gathercole, George E. Deputy minister of economics 1956–62, first vice-chairman of the HEPC 1961–6, chairman of the HEPC 1966–74, and chairman of Ontario Hydro 1974. Correspondence, 21 May and 17 August 1992.

Gordon, Douglas J. General manager of the HEPC 1970–4, member of Task Force Hydro 1971–3, and president of Ontario Hydro 1974–80. Interviews, Willowdale, Ontario, 15 November 1991 and 26 June 1992.

Jennings, I.H. Tony. Chief executive officer, Municipal Electric Association, Toronto, Ontario. Correspondence, 17 August 1992.

Leonoff, Lawrence E. Vice-president, general counsel and secretary, Ontario Hydro. Correspondence, 17 October 1991 and 24 August 1992.

Macaulay, Hon. Robert W. HEPC commissioner 1958–63, minister of energy resources 1959–63, and chairman of the Ontario Energy Board 1984–7. Interviews, Toronto, Ontario, 30 July 1991, 31 October 1991 (by telephone), and 23 June 1992.

Macdonald, H. Ian. Provincial economist 1965, deputy treasurer of Ontario 1965–74, member of the Committee on Government Productivity 1969–73, and member of the Advisory Committee on Energy 1971–3. Interview, North York, Ontario, 26 June 1992.

McDowall, Duncan. Professor of history, Carleton University, Ottawa, Ontario. Correspondence, 21 September 1992.

McHenry, Gordon M. Formerly general manager of personnel, Ontario Hydro. Interview, Islington, Ontario, 17 October 1991.

McKeough, Hon. W. Darcy. Minister of municipal affairs 1967–71, treasurer of Ontario 1971–2, parliamentary assistant to the premier 1973, minister of energy 1973–5, and treasurer of Ontario 1975–8. Interview, Toronto, Ontario, 6 May 1992.

Muncaster, J. Dean. Formerly chairman of Task Force Hydro 1971–3, HEPC commissioner 1973–4, and director of Ontario Hydro 1974–80. Interview, Toronto, Ontario, 12 May 1992.

Nelles, H.V. Professor of history, York University, North York, Ontario. Correspondence, undated [October 1990] [quoted with permission].

Nokes, E.C. Secretary-manager, Ontario Municipal Electrical Association, 1965–85, and chief executive officer, Municipal Electric Association, 1985–6. Interview, St Catharines, Ontario, 22 July 1994 (by telephone).

Reynolds, Dr J. Keith. Deputy minister to the prime minister of Ontario 1964–72, secretary to the cabinet 1969–72, member of the Committee on Government Productivity 1969–73, member of Task Force Hydro 1971–3, member of the Advisory Committee on Energy 1971–3, and deputy provincial secretary for resources development 1972–4. Interview, Scarborough, Ontario, 9 March 1992.

Strike, Alan. Son of W.R. Strike (president of the Ontario Municipal Electric Association 1944–6, HEPC commissioner 1944–61, HEPC chairman 1961–6). Interview, Bowmanville, Ontario, 18 October 1991.

Taylor, Robert B. Member of Task Force Hydro 1971–3, vice-chairman of Ontario Hydro 1974, and chairman of Ontario Hydro 1975–9. Interview, Willowdale, Ontario, 18 June 1992.

Books, Articles, and Documents

Advisory Committee on Energy. *Energy in Ontario, The Outlook and Policy Implication: Volume One*. 19 December 1972.

– *Energy in Ontario, The Outlook and Policy Implications: Volume Two*. 5 March 1973.

Armstrong, Christopher. *The Politics of Federalism: Ontario's Relations with the Federal Government, 1867–1940*. Toronto: Ontario Historical Studies Series/University of Toronto Press, 1981.

– and H.V. Nelles. *Monopoly's Moment: The Organization and Regulation of Canadian Utilities, 1830–1930*. Toronto: University of Toronto Press, 1988.

Atomic Energy of Canada Limited. *Annual Report, 1952–3*. Ottawa, 1953.

Black, Edwin R., and Alan C. Cairns. 'A Different Perspective on Canadian Federalism.' *Canadian Public Administration* 9, no. 1 (Spring 1966).

Blake, Cassels & Graydon. Legal opinions for Task Force Hydro, 13 June 1972 and 19 September 1972.

Bolton, Reginald Pelham. *An Expensive Experiment: The Hydro-Electric Power Commission of Ontario*. New York: The Baker and Taylor Company, 1913.

Bothwell, Robert. *Eldorado: Canada's National Uranium Company*. Toronto: University of Toronto Press, 1984.

– *Loring Christie and the Failure of Bureaucratic Imperialism*. New York: Garland Publishing, 1988.

– *Nucleus: The History of Atomic Energy of Canada Limited*. Toronto: University of Toronto Press, 1988.

Brady, Alexander. 'The Ontario Hydro-Electric Power Commission.' *Canadian Journal of Economics and Political Science* 2, no. 3 (1936).
Canada, *Senate Debates, 1932–3.*
Canadian Annual Review, 1902–38 and 1974–87. Toronto: The Canadian Annual Review Publishing Company, 1903–17; Toronto: The Canadian Annual Review Limited, 1918–20; Toronto: The Canadian Review Company Limited, 1921–40; Toronto: University of Toronto Press, 1960–87.
Canadian Parliamentary Guide, 1903–73.
Canadian Who's Who, 1910 and 1936–94. London: Times Publishing Company, 1910; Toronto: Trans-Canada Press, 1938–66; Toronto: Who's Who Canadian Publications, 1967–75; Toronto: University of Toronto Press, 1979–94.
Chandler, Marsha A. 'The Politics of Public Enterprise,' in *Crown Corporations in Canada: The Calculus of Instrument Choice*. Edited by J.R.S. Prichard. Toronto: Butterworths, 1983.
– and William Chandler, *Public Policy and Provincial Politics*. Toronto: McGraw-Hill Ryerson, 1979.
City of Toronto. *Minutes of the Proceedings of the Council of the Corporation of the City of Toronto*, 1907.
Committee on Government Productivity. *Interim Report Number One*. Toronto: Queen's Printer, 15 December 1970.
– *Interim Report Number Two*. Toronto: Queen's Printer, 16 March 1971.
– *Interim Report Number Three*. Toronto: Queen's Printer, December 1971.
– *Interim Report Number Four*. Toronto: Queen's Printer, 22 December 1971.
– *Report Number Nine*. Toronto: COGP, March 1973.
– *Report Number Ten*. Toronto: COGP, March 1973.
Cook, Peter. *Massey on the Brink: The Story of Canada's Great Multinational and Its Struggle to Survive*. Toronto: Collins, 1981.
Cooke, J.R. *Hydro Service Considered in Some of Its Important Economic Aspects*. Liberal-Conservative Summer School, September 1933.
Dales, John H. *Hydro-Electricity and Industrialization: Quebec, 1898 to 1940*. Cambridge, Mass.: Harvard University Press, 1957.
Denison, Merrill. *The People's Power: The History of Ontario Hydro*. Toronto: McClelland and Stewart, 1960.
Dewar, Kenneth C. 'Private Electrical Utilities and Municipal Ownership in Ontario, 1891–1900.' *Urban History Review* 12, no. 1 (June 1983).
– 'State Ownership in Canada: The Origins of Ontario Hydro.' Ph.D. thesis, University of Toronto, 1975.
Doern, G. Bruce. *Government Intervention in the Canadian Nuclear Industry*. Montreal: The Institute for Research on Public Policy, 1980.

Duncan, James S. *Brief to the Committee on the Organization of Government in Ontario.* HEPC: Toronto, 18 November 1958.
Eggleston, Wilfrid. *Canada's Nuclear Story.* Toronto and Vancouver: Clarke, Irwin and Company, 1965.
Fleming, Keith R. *Power at Cost: Ontario Hydro and Rural Electrification, 1911–58.* Montreal and Kingston: McGill-Queen's University Press, 1992.
Forman, Debra. *Legislators and Legislatures of Ontario: A Reference Guide*, 4 vols. Toronto: Legislative Library, 1984 and 1991.
Freeman, Neil B. 'Ontario Hydro and Its Government, 1906–73: A Contemporary Analysis of Its Historical Relationship.' Ph.D. thesis, University of Toronto, 1993.
– 'Turn-of-the-Century State Intervention: Creation of the Hydro-Electric Power Commission of Ontario, 1906.' *Ontario History* 84, no. 3 (September 1992).
Gaby, F.A. *Some Interesting Aspects of the Hydro System.* Toronto: HEPC, 1931.
– *Trends of Electrical Demands in Relation to Power Supplies.* Toronto: HEPC, 1933.
– *The Power Situation in Ontario.* Toronto: Power Securities Protective Association, 1935.
Gagnon, Georgette, and Dan Rath. *Not without Cause: David Peterson's Fall from Grace.* Toronto: HarperCollins, 1991.
Gordon, Walter L. *The Hydro-Electric Power Commission of Ontario: Report upon the Proposed Organization Plan.* Toronto: J.D. Woods and Gordon Limited, 24 March 1947.
Graham, Roger. *Arthur Meighen, Volume III: No Surrender.* Toronto: Clarke, Irwin, 1965.
– *Old Man Ontario: Leslie M. Frost.* Toronto: Ontario Historical Studies Series/ University of Toronto Press, 1990.
Heard, Andrew. *Canadian Constitutional Conventions: The Marriage of Law and Politics.* Toronto: Oxford University Press, 1991.
HEPC. *Annual Reports*, 1908–73. Toronto: HEPC, 1910–74.
– *Errors and Misrepresentations made by the Hydro-Electric Inquiry Commission (Known as the Gregory Commission) respecting the Publicly Owned and Operated Hydro-Electric Power Undertaking of the Municipalities of the Province of Ontario.* Toronto, HEPC, 1925.
– *Hydro News*, vols. 29–61 (1942–74).
– *Interim Report on a Proposal to Standardize Frequency at 60 Cycles for the Southern Ontario System.* Toronto: HEPC, November 1946.
– *Misleading Assertions that have been made relating to the Power Situation in the Province of Ontario Examined and Corrected.* Toronto: HEPC, n.d. [22 August 1933].
– *Misleading Assertions that have been made relating to the Power Situation in the Province of Ontario Have Not Been Withdrawn.* Toronto: HEPC, 31 August 1933.

– *Paid-for Propaganda???: Who Instigates Attacks on Hydro? Important Facts Brought to Public Attention by the Hydro-Electric Power Commission of Ontario*. Toronto: HEPC, 7 June 1934.
– *Refutation of Unjust Statements contained in a report published by the National Electric Light Association entitled 'Government Owned and Controlled Compared with Privately Owned and Regulated Electric Utilities in Canada and the United States' respecting the Hydro-Electric Power Commission of Ontario*. Toronto: HEPC, 1922.
– *Statement Respecting Findings and other Statements contained in the Majority Report of the Sutherland Commission*. Toronto: HEPC, 1922.
– *Statistical Yearbook, 1971*. Toronto: HEPC [1972].
– *Summary of the Reports of the Commission's Consultants concerning the Problem of Frequency Standardization in the Southern Ontario System of the Hydro-Electric Power Commission of Ontario*. Toronto: HEPC, 1 March 1948.
– *The Bulletin*, vols. 1–29 (1914–42).

Hodgetts, J.E. *Administering the Atom for Peace*. New York: Atherton Press, 1964.
– *From Arm's Length to Hands-On: The Formative Years of Ontario's Public Service, 1867–1940*. Toronto: Ontario Historical Studies Series/University of Toronto Press, 1995.

Hogg, T.H. 'Power Resources and Requirements in relation to the Quebec Power Contracts Settlement.' Address to the OMEA, 8 February 1937, reprinted in HEPC, *The Bulletin* 25, no. 2 (February 1938).
– 'Mutual Interdependence.' Address to the OMEA 4 July 1939, reprinted in HEPC, *The Bulletin* 26, no. 7 (July 1939).
– 'Hydro's Post-war Challenges.' Address to the OMEA, 5 March 1946 (unpublished, although enclosed in HEPC, *Hydro News* (April 1946).

Holland, Stuart. 'Europe's New Public Enterprises,' in *Big Business and the State*. Edited by Raymond Vernon. Cambridge, Mass.: Harvard University Press, 1974.

Houck, W.L. 'The Municipalities and the Hydro Scheme.' Address to the OMEA, 8 February 1938, reprinted in HEPC, *The Bulletin* 25, no. 2 (February 1938).

Humphries, Charles W. *'Honest Enough to Be Bold': The Life and Times of Sir James Pliny Whitney.* Toronto: Ontario Historical Studies Series/University of Toronto Press, 1985.

Hutcheson, George F. *Head and Tales: Memoirs of George F. Hutcheson*. Bracebridge: Herald-Gazette Press, 1972.

Johnson, Charles M. *E.C. Drury: Agrarian Idealist.* Toronto: Ontario Historical Studies Series/University of Toronto Press, 1986.

Jones, Murray V. *Final Report and Recommendations: Ottawa, Eastview and Carleton County Local Government Review.* June 1965.

Judicial Committee of the Privy Council. *St Catharines* v. *Hydro-Electric Power Commission* [1929]. Dominion Law Reports, 1930.

Laux, Jean Kirk, and Maureen Appel Molot. *State Capitalism: Public Enterprise in Canada*. Ithaca, NY: Cornell University Press, 1988.

Legislature of Ontario. *Debates*, 16 April 1948; 15 March 1922; 10 February 1959; 28 March 1961; 22 July 1971; 30 March 1972; 30 April 1973; and 7 June 1973.

Macaulay, Robert W. *A Layman's Guide to Nuclear Power*. 1959.

– *The World of Energy*. Toronto: Newton, 1961.

MacDonald, Donald C., ed. *Government and Politics in Ontario*. Toronto: Macmillan, 1975.

– *The Happy Warrior: Political Memoirs*. Markham, Ont.: Fitzhenry and Whiteside, 1988.

Manthorpe, Jonathan. *Power and the Tories: Ontario Politics, 1943 to the Present*. Toronto: Macmillan, 1974.

Mavor, James. *Public Ownership and the Hydro-Electric Power Commission of Ontario*. Toronto: Maclean Publishing Company, 1917.

– *Niagara in Politics: A Critical Account of the Ontario Hydro-Electric Commission*. New York: E.P. Dutton and Company, 1925.

McCrimmon, A. Murray. *Report to the Chairman [of the HEPC by the Assistant Secretary and Controller]* (private and confidential) 7 August 1935. Archives of Ontario, MU 2496, Roebuck Papers, general, May-August 1935.

McDougall, A.K. *John P. Robarts: His Life and Government*. Toronto: Ontario Historical Studies Series/University of Toronto Press, 1986.

McDowall, Duncan. *The Light: Brazilian Traction, Light and Power, 1899–1945*. Toronto: University of Toronto Press, 1988.

McKenty, Neil. *Mitch Hepburn*. Toronto: McClelland and Stewart, 1967.

McKeough, W. Darcy. *Report on the Structure of Energy Administration in Ontario*. 1 June 1973.

Morrison, Herbert. *Government and Parliament: A Survey from the Inside*. London: Oxford University Press, 1954.

Municipal Electric Association. *Restructuring the Electricity Industry in Ontario, Volume I: Recommended Strategy: Final Phase III Report of the Ad Hoc Task Force to Review Institutional Options for Electricity in Ontario*. Toronto: Municipal Electric Association, 6 September 1994.

– *Phase I Report of the Ad Hoc Task Force to Review Institutional Options for Electricity in Ontario: A Discussion Paper*. Toronto: Municipal Electric Association, 27 May 1993.

– *Phase II Report of the Ad Hoc Task Force to Review Institutional Options for Electricity in Ontario: A Discussion Paper*. Toronto: Municipal Electric Association, 19 August 1993.

– *Phase III: Interim Report of the Ad Hoc Task Force to Review Institutional Options for Electricity in Ontario: A Discussion Paper*. Toronto: Municipal Electric Association, 31 March 1994.

Murray, W.S. *Government Owned and Controlled Compared with Privately Owned and Regulated Electric Utilities in Canada and the United States*. New York: National Electric Light Association, 1922.

Neatby, H. Blair. *William Lyon Mackenzie King, 1924–32: The Lonely Heights*. Toronto: University of Toronto Press, 1963.

Nelles, H.V. *The Politics of Development: Forests, Mines and Hydro-Electric Power in Ontario, 1849–1941*. Toronto: Macmillan, 1974.

Oliver, Peter. *G. Howard Ferguson: Ontario Tory*. Toronto: Ontario Historical Studies Series/University of Toronto Press, 1977.

– 'Sir William Hearst and the Collapse of the Conservative Party,' in *Public and Private Persons: The Ontario Political Culture, 1914–34*. Edited by Peter Oliver. Toronto: Clarke, Irwin and Company, 1975.

Ontario. *Bills of Ontario*, 1906, 1959 and 1960–1.

– *Design for Development [Phase One]*. Statement by the Prime Minister of the Province of Ontario on Regional Development Policy, 5 April 1966.

– *Design for Development: Phase Two*. Statement by the Honourable John Robarts, Prime Minister of Ontario, 28 November 1968, and Statement by the Honourable W. Darcy McKeough, Minister of Municipal Affairs, 2 December 1968.

– *Design for Development: Phase Three*. Statement by the Honourable William G. Davis, Premier of Ontario, to the Legislature, 16 June 1972, and Statement by the Honourable W. Darcy McKeough, Treasurer of Ontario, to the Founding Convention of the Association of Municipalities of Ontario, Ottawa, 19 June 1972.

– *Final Report of the Select Committee of the Legislature Investigating Ontario Hydro: A New Public Policy Direction for Ontario Hydro*. Toronto: 18 June 1976.

– *Fourth and Final Report of the Select Committee on the Municipal and Related Acts*. March 1965 (Beckett Report).

– *Hydro-Electric Inquiry Commission: General Report*. Toronto: 5 March 1924 (Gregory Report).

– *Hydro-Electric Inquiry Commission: Report on History and General Relations*. 23 December 1923.

– *Hydro-Electric Power Commission of the Province of Ontario: First Report; Niagara District*. 4 April 1906 (Beck Report).

– *Interim Report of the Select Committee of the House to Examine into and Study the Administrative and Executive Problems of the Government of Ontario*. Toronto: 17 November 1960 (Roberts Report).

– *Journals of the Legislative Assembly of the Province of Ontario*. Toronto: King's Printer 1911, 1916, 1933, 1935, 1939, 1948; Queen's Printer, 1959, 1960–1, 1961–2, 1964, 1968, 1973, 1987–9, and 1990–1.

– *Ontario Commission on the Legislature*, 5 vols., 1973–5 (Camp Report).

– 'Public Accounts, 1914–15.' *Sessional Papers, 1916*.

– *Report of the Commission Appointed to Inquire into Hydro-Electric Radials*. Toronto, 1921 (Sutherland Report).
– 'Report of the Committee on Public Accounts, 1916.' Appendix One, *Journals of the Legislative Assembly of the Province of Ontario, 1916*.
– *Report of the Committee on the Organization of Government in Ontario*. Toronto: 25 September 1959 (Gordon Report).
– *Report of the Royal Commission on Electric Power Planning: Volume 1; Concepts, Conclusions, and Recommendations*. Toronto: 28 February 1980 (Porter Report).
– *Report of the Select Committee on the Hydro-Electric Power Commission New Head Office Building, 1973*. Toronto: 2 October 1973 (MacBeth Report).
– *Response of the Government Ontario to the Royal Commission on Electric Power Planning*. May 1981.
– *Royal Commission Appointed to Inquire into Certain Matters Concerning the Hydro-Electric Power Commission of Ontario*. Toronto: 31 October 1932.
– *Royal Commission Appointed to Inquire into the Purchase of Bonds of the Ontario Power Service Corporation by the Hydro-Electric Power Commission of Ontario and the Government of Ontario, and the Payment therefore in the Bonds of the Hydro-Electric Power Commission of Ontario, and all circumstances connected therewith*. Toronto: 20 October 1934 (Latchford-Smith Report). Printed as 'Royal Commission re: Abitibi Inquiry: Report of Commissioners' in HEPC, *Annual Report, 1934*.
– *Select Committee on Ontario Hydro Affairs: Report on Proposed Uranium Contracts*. March 1978.
– 'Select Committee to Investigate Power Contracts with Quebec Power Companies and Other Hydro Matters.' Appendix 1 of Ontario, *Journals of the Legislative Assembly of the Province of Ontario, 1939*.
– 'Statement by the Honourable Allan Grossman, Provincial Secretary for Resources Development, on the Long-range Planning of Ontario's Electric Power System.' 13 March 1975.
– 'Statement by the Honourable William G. Davis, Premier of Ontario, at a News Conference to release the First Report of Task Force Hydro.' 13 November 1972.
– *Taxation in Ontario: A Program of Reform: The Report of the Select Committee of the Legislature on the Report of the Ontario Committee on Taxation*. Toronto: 16 September 1968 (White Report).
– *Third Report of the Select Committee on the Municipal and Related Acts*. March 1964 (Beckett Report).
Ontario Committee on Taxation. 'Local Non-Tax Revenue,' in *Report II: The Local Revenue System*. Toronto: Queen's Printer, 1967 (Smith Report).
– 'Provincial Government Enterprises,' in *Report III: The Provincial Revenue System*. Toronto: Queen's Printer, 1967 (Smith Report).

– 'Reconciling Structure with Finance,' in *Report II: The Local Revenue System.* Toronto: Queen's Printer, 1967 (Smith Report).

Ontario Energy Board. *Report of the Board: Reference re. International Activities of Ontario Hydro International Incorporated.* Toronto: 10 April 1995.

Ontario Hydro. *Annual Report*, 1974–94.

– *Hydro 21: Options for Ontario Hydro – Present Situation and Options for the Future.* Toronto: Ontario Hydro, February 1993.

– *Hydro in Transition: Employee Information.* Toronto: Ontario Hydro, 9 March 1993.

– *Ontario Hydro and the Electric Power Industry: Challenges and Choices [Working Draft of Report of the Financial Restructuring Group].* Toronto: Ontario Hydro, 27 June 1994.

– *Providing the Balance of Power: Ontario Hydro's Plan to Service Customers' Electricity Needs.* Toronto: Ontario Hydro, December 1989.

– *The New Ontario Hydro.* Toronto: Ontario Hydro, 14 April 1993.

Ontario Municipal Electric Association. *Annual Meeting*, 1930–48, 1952–66, and *Annual Meeting* full proceedings, 1967–74.

– *Minutes of the Executive Committee, 1948.*

– *Power Costing in Ontario.* n.d. [1971].

– *Special General Meeting, 1948.*

– *Submission to the Premier of the Province of Ontario with respect to Task Force Hydro's Report to the Executive Council on Hydro in Ontario – A Future Role and Place.* 19 September 1972. Archives of Ontario, RG 18, E-23, Box 8, File 1.

– *Summer Convention*, 1931–9.

– *The OMEA – Ontario Hydro, the People's Power – and Government.* Toronto: OMEA, February 1963.

Ontario Power Commission. *Official Report of the Ontario Power Commission.* Toronto: Monetary Times Publishing Company, 28 March 1906 (Snider Report).

Plewman, W.R. *Adam Beck and the Ontario Hydro.* Toronto: Ryerson, 1947.

Prang, Margaret. *N.W. Rowell: Ontario Nationalist.* Toronto: University of Toronto Press, 1975.

Progressive Conservative Party of Ontario. *The Common Sense Revolution.* n.d. [1994].

Regehr, T.D. *The Beauharnois Scandal: A Story in Canadian Entrepreneurship and Politics.* Toronto: University of Toronto Press, 1990.

Roebuck, Arthur W. *The Wreck of the Hydro.* Toronto: Ontario Liberal Association, 19 August 1933.

– 'Crisis in the Affairs of Hydro.' Four addresses to the legislature, 26–29 March 1935. Reprinted in HEPC, *The Bulletin* 22, no. 5 (May 1935).

– 'Exhaustive Summary of Hydro Affairs.' Address to the legislature, 3 and 4 March 1936. Reprinted in HEPC, *The Bulletin* 23, no. 3 (March 1936).

Saywell, John T. *'Just Call Me Mitch': The Life of Mitchell F. Hepburn*. Toronto: Ontario Historical Studies Series/University of Toronto Press, 1991.

Schindeler, F.F. *Responsible Government in Ontario*. Toronto: University of Toronto Press, 1969.

Silcox, Peter. 'The ABC's of Ontario: Provincial Agencies, Boards and Commissions,' in *Government and Politics in Ontario*. Edited by Donald C. MacDonald. Toronto: Macmillan, 1975.

Solandt Commission. *Interim Report*. 31 October 1972.

– *'Closing the Gap': A Public Inquiry into the Transmission of Power between Nanticoke and Pickering*. March 1974.

– *'Transmission': A Public Inquiry into the Transmission of Power between Lennox and Oshawa*. April 1975.

Speirs, Rosemary. *Out of the Blue: The Fall of the Tory Dynasty in Ontario*. Toronto: Macmillan, 1986.

State of New York. *Report of the Joint Committee of the Legislature on the Conservation of Water*. 30 January 1912 (Ferris Report, 1912).

– *Report of the Joint Committee of the Legislature on the Conservation and Utilization of Water Power*. 15 January 1913 (Ferris Report, 1913).

Strong, Maurice. 'Hydro in Ontario: Strengthening the Partnership.' Speech to the 1993 MEA annual meeting, 1 March 1993.

– 'Hydro in Ontario: Forging a New Partnership.' Speech to the 1994 MEA annual meeting, 28 February 1994.

Supreme Court of Canada. *Hydro-Electric Commission of the Township of Nepean* v. *Ontario Hydro*. Dominion Law Reports, 1982.

Szablowski, George J. 'Policy-Making in the Cabinet: Recent Organizational Engineering at Queen's Park,' in *Government and Politics in Ontario*. Edited by Donald C. MacDonald. Toronto: Macmillan, 1975.

Task Force Hydro. *Report Number One: Hydro in Ontario, A Future Role and Place*. 15 August 1972.

– *Report Number Two: Hydro in Ontario, An Approach to Organization*. 14 December 1972.

– *Report Number Three: Nuclear Power in Ontario*. 16 February 1973.

– *Report Number Four: Hydro in Ontario, Financial Policy and Rates*. 11 April 1973.

– *Report Number Five: Hydro in Ontario; A Policy for Make or Buy*. 29 June 1973.

– *Response to OMEA Statements on Task Force Hydro Report Number One*. 10 January 1973. Archives of Ontario, R.G. 18, E–23, Box 8, File 25.

Tennyson, Brian. 'Sir Adam Beck and the Ontario General Election of 1919.' *Ontario History* 45, no. 2 (Fall 1966).

Trebilcock, M.J., J.R.S. Prichard, D.G. Hartle, and D.N. Dewees. *The Choice of Governing Instrument*. Ottawa: Economic Council of Canada, 1982.

Tuohy, Carolyn J. *Policy and Politics in Canada: Institutionalized Ambivalence*. Philadelphia: Temple University Press, 1992.

Walkom, Thomas. *Rae Days: The Rise and Follies of the NDP.* Toronto: Key Porter Books, 1994.

White, Graham. *The Ontario Legislature: A Political Analysis*. Toronto: University of Toronto Press, 1989.

Young, R.A., Philippe Faucher, and André Blais. 'The Concept of Province-Building: A Critique.' *Canadian Journal of Political Science* 17, no. 4 (December 1984).

Index